Aquaculture Technology
Flowing Water and Static Water Fish Culture

T0199512

Aquaculture Technology

Flowing Water and Static Water Fish Culture

Richard W. Soderberg
Mansfield University
Mansfield, Pennsylvania

CRC Press
Taylor & Francis Group
Boca Raton London New York

CRC Press is an imprint of the
Taylor & Francis Group, an **informa** business

CRC Press
Taylor & Francis Group
6000 Broken Sound Parkway NW, Suite 300
Boca Raton, FL 33487-2742

First issued in paperback 2020

ISBN 13: 978-0-367-57374-4 (pbk)
ISBN 13: 978-1-4987-9884-6 (hbk)

Library of Congress Cataloging-in-Publication Data

Names: Soderberg, Richard W., author.
Title: Aquaculture technology : flowing water and static water fish culture / Richard W. Soderberg.
Description: Boca Raton : Taylor & Francis, 2017. | Includes bibliographical references and index.
Identifiers: LCCN 2016050243 | ISBN 9781498798846 (hardback : alk. paper)
Subjects: LCSH: Fish culture--Water-supply. | Aquaculture.
Classification: LCC SH154 .S63 2017 | DDC 639.8--dc23
LC record available at https://lccn.loc.gov/2016050243

Visit the Taylor & Francis Web site at
http://www.taylorandfrancis.com

and the CRC Press Web site at
http://www.crcpress.com

Visit the eResources: https://www.crcpress.com/9781498798846

Contents

Preface

The science of aquaculture is best divided into two general components: flowing water and static water technologies. In flowing water fish culture systems, a constant stream of water carries dissolved oxygen to the fish and flushes waste materials away from them. Water temperature, which controls fish growth rates, is constant or nearly so. In static water systems, dissolved oxygen dynamics are controlled by the dynamic processes of photosynthesis and respiration, and waste materials are metabolized by biological and chemical processes within the pool of stagnant water. Temperatures of surface waters vary daily and seasonally so that ambient air temperatures determine species selection and fish growth rates.

This volume was designed as the text for a two-semester college course in aquaculture. Section I, "Flowing Water Fish Culture," describes the simpler technology and is appropriate for the first-semester course. The applied limnology of static water fish culture, Section II, is intended to serve as the second-semester text. Students using this book as a college text should have a basic background in mathematics, chemistry, and limnology. I have included a review of the basic principles of limnology for curricula that do not include that subject.

My first book, *Flowing Water Fish Culture*, published by Lewis Publishers in 1995, has seen widespread use by aquaculture practitioners and has been used as a college text, and this two-section volume is intended to serve the wider aquaculture community as well.

Completion of this book would not have been possible without constructive criticism from my colleagues and students at Auburn University and Mansfield University. I especially thank Professor Claude E. Boyd for so eloquently demonstrating that the solutions to biological and chemical problems lie in the quantitative investigation of their components.

Richard W. Soderberg
Mansfield, Pennsylvania

Flowing water fish culture

chapter one

Flowing water fish culture

The practice of flowing water fish culture in the United States began in the mid-1800s in response to dwindling numbers of native brook trout. The facilities were little more than fenced-off portions of streams resembling trout habitat. Eventually, all states with trout fisheries established fishery agencies and built hatcheries, and the artificial reproduction and husbandry of salmonids was documented.

There are two basic considerations for the controlled growth of fish in a stream of water. First, the medium that supplies oxygen for respiration and flushes away metabolic waste is water, which has a low affinity for oxygen and occurs in finite quantities. This problem becomes obvious when aquaculture is compared to terrestrial animal husbandry in air, which contains a great deal of oxygen and is available in unlimited supply. Second, fish are cold-blooded and only grow satisfactorily within rather narrow temperature ranges. Thus, water temperature determines which species, if any, may be produced. Warm-blooded animals, on the other hand, use food energy to maintain their optimum growth temperatures regardless of environmental temperature.

In the 1950s, David Haskell, an engineer employed by the New York State Department of Environmental Conservation, first applied analytical investigation to the art of flowing water fish culture. Haskell's quantitative approach to the definition of chemical and biological parameters affecting fish in confinement allowed fish culture to progress from an art to a science.

Haskell's pioneering work resulted in the elucidation of five basic principles upon which our present understanding of flowing water fish culture is based.

1. At constant temperature, fish growth, in units of length, is linear over time until sexual maturity is approached.
2. The growth rate of fish, in units of length, is proportional to temperature. Therefore, if the growth rate at one temperature is known, the growth rate at another temperature may be predicted.
3. Feeding rates can be rationally predicted based on estimated food conversion, metabolic characteristics, and the anticipated growth rate.

4. The maximum permissible weight of fish that can be supported in a rearing unit is determined by the depletion of oxygen and the accumulation of metabolic wastes.
5. Oxygen consumption and metabolite production progress in proportion to the amount of food fed.

Based upon this framework, flowing water fish culture has become a quantitative agricultural science. The technology of that science is the subject of the chapters that comprise Section I of this book.

chapter two

Fish growth in hatcheries

Growth rates of cultured fish in relation to temperature

Fish growth models based on temperature units

Haskell (1959) plotted fish growth rate against temperature in the range of 42–52°F (5.5–11.1°C) and found that they were linearly related with an intercept at 38.6°F (3.7°C) (Figure 2.1).

Growth does not cease at 38.6°F, but the intercept is necessary to establish the line from which growth is predicted. Haskell (1959) defined a temperature unit (TU) as 1°F over 38.6°F for 1 month. For example, if the monthly water temperature averaged 50°F, 50–38.6°F = 11.4 TU would be available during that month. Haskell reported that approximately 21 TUs were required per inch of trout growth in New York hatcheries. In an experiment in which brook trout (*Salvelinus fontinalis*) were grown under conditions of "optimum care," 16.2 TUs were required per inch of growth (Haskell 1959), but Haskell doubted that trout could grow that fast under hatchery conditions. Modern hatcheries experience growth rates of approximately 17 TU/inch for brook, brown (*Salmo trutta*), and rainbow trout (*Oncorhynchus mykiss*), possibly due to improvements in hatchery management and fish diets that have occurred since Haskell's work. The actual growth rate, in temperature units required per length increment of growth, should be determined for each hatchery, species, and strain of fish.

The U.S. Fish and Wildlife Service (Piper et al. 1982) has adopted monthly temperature units (MTU), defined as the average monthly temperature in centigrade minus the freezing point of water. Therefore, the MTU theory differs from Haskell's TU model by assuming linear growth at temperatures cooler than 42°F (5.5°C). This theory forces the growth line through the origin, which could result in substantial error in growth-rate estimation, especially for warmwater fish whose zero-growth intercepts are at higher temperatures than those of salmonids. In Figure 2.2, Haskell's (1959) data for brook trout are shown on a growth versus temperature plot containing a line of best fit for the data when it is forced through the origin of the graph. It is evident that ignoring the zero-growth intercept compromises the accuracy of the model.

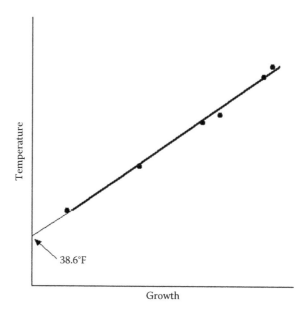

Figure 2.1 Plot of trout growth in units of fish length at temperatures from 42 to 52°F. (Data from Haskell, D. C. 1959. *New York Fish and Game Journal* 6: 205–237.)

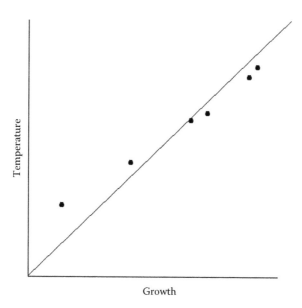

Figure 2.2 When Haskell's (1959) data are fitted to a regression line forced through the origin of the graph in accordance with the MTU growth prediction method (Piper et al. 1982), the accuracy of the model is compromised.

Table 2.1 MTU (average monthly temperature, °C) required per centimeter of brook trout growth at temperatures from 4 to 19°C

Exposure temperature	MTU required/cm of growth
4	15.2a
7	8.7b
10	5.8bc
13	5.7bc
16	6.8bc
19	9.1b
Mean (7–19°C)	7.2
Mean (10–16°C)	6.1

Source: After Dwyer, W. P., R. G. Piper, and C. E. Smith. 1981a. *Brook Trout Growth Efficiency as Affected by Temperature.* Information Leaflet 16, U.S. Fish and Wildlife Service Fish Culture Development Center, Bozeman, Montana.

Note: Values followed by the same letter are not significantly different (P > 0.05).

Research at the U.S. Fish and Wildlife Service Bozeman Fish Technology Center in Montana has defined MTU requirements for some salmonid species. Brook trout (Dwyer et al. 1981a) required 7.2 MTU/ centimeter (cm) of growth in the temperature range of 7–19°C, but only 6.1 MTU/cm from 10 to 16°C (Table 2.1).

Rainbow trout (Dwyer et al. 1981b) required 5.8 MTU/cm at 13°C and an average of 7.1 MTU/cm in the temperature range of 7–19°C (Table 2.2).

Table 2.2 MTU (average monthly temperature, °C) required per centimeter of rainbow trout growth at temperatures from 4 to 19°C

Exposure temperature	MTU required/cm of growth
4	12.4a
7	8.7bd
10	6.6bcd
13	5.8b
16	6.3bcd
19	8.2bcd
Mean (7–19°C)	7.1

Source: After Dwyer, W. P., R. G. Piper, and C. E. Smith. 1981b. *Rainbow Trout Growth Efficiency as Affected by Temperature.* Information Leaflet 18, U.S. Fish and Wildlife Service Fish Culture Development Center, Bozeman, Montana.

Note: Values followed by the same letter are not significantly different (P > 0.05).

Table 2.3 MTU (average monthly temperature, °C) required per centimeter of lake trout growth at temperatures from 4 to 16°C

Exposure temperature	MTU required/cm of growth
4	5.0a
7	4.5a
10	5.1a
13	5.4a
16	7.0a
Mean	5.4

Source: After Dwyer, W. P., R. G. Piper, and C. E. Smith. 1981c. *Lake Trout, Salvelinus namaycush, Growth Efficiency as Affected by Temperature.* Information Leaflet 22, U.S. Fish and Wildlife Service Fish Culture Development Center, Bozeman, Montana.

Note: Values followed by the same letter are not significantly different (P > 0.05).

In the temperature range of 4–16°C, lake trout (*Salvelinus namaycush*) (Dwyer et al. 1981c) required 5.4 MTU/cm of growth (Table 2.3).

Steelhead trout (*Oncorhynchus mykiss*) (Dwyer et al. 1982) required 10.9 MTU/cm in the range from 4 to 19°C, but only 8.5 MTU/cm at 13°C (Table 2.4).

An average of 10.1 MTU/cm were required for Atlantic salmon (*Salmo salar*) (Dwyer and Piper 1987) at temperatures ranging from 7 to 16°C, but growth efficiency was significantly reduced at 4 and 19°C (Table 2.5).

Table 2.4 MTU (average monthly temperature, °C) required per centimeter of steelhead trout growth at temperatures from 4 to 19°C

Exposure temperature	MTU required/cm of growth
4	15.0a
7	9.6a
10	9.5a
13	8.5b
16	10.0a
19	12.7a
Mean	10.9

Source: After Dwyer, W. P., R. G. Piper, and C. E. Smith. 1982. *Steelhead Trout Growth Efficiency as Affected by Temperature.* Information Leaflet 27, U.S. Fish and Wildlife Service Fish Culture Development Center, Bozeman, Montana.

Note: Values followed by the same letter are not significantly different (P > 0.05).

Table 2.5 MTU (average monthly temperature, °C) required per centimeter of Atlantic salmon growth at temperatures from 4 to 19°C

Exposure temperature	MTU required/cm of growth
4	14.6a
7	10.9b
10	10.0b
13	9.9b
16	9.7b
19	124a
Mean (7–16°C)	10.1

Source: After Dwyer, W. P. and R. G. Piper. 1987. *Progressive Fish-Culturist* 49: 57–59.

Note: Values followed by the same letter are not significantly different ($P > 0.05$).

Notice that the temperature units required per cm of growth for these species are generally greater at the extremes of the temperature ranges tested than within a narrower, optimum temperature range. This demonstrates that fish grow most efficiently within rather narrow ranges of temperature.

Andrews et al. (1972) presented data on the growth of channel catfish (*Ictalurus punctatus*) at temperatures from 24 to 30°C. When their reported final fish weights are converted to length (Piper et al. 1982) and length increment is plotted against temperature, the intercept is negative. Thus, the TU procedure is inappropriate for these data. When the MTU method of growth projection is applied, a usable, albeit compromised, growth model for catfish is obtained (Table 2.6).

Table 2.6 MTU (average monthly temperature, °C) required per centimeter of channel catfish growth at temperatures from 24 to 30°C

Exposure temperature	MTU required/cm of growth
24	6.1
26	6.1
28	6.3
30	6.8
Mean	6.3

Source: Data from Andrews, J. W., L. H. Knight, and T. Murai. 1972. *Progressive Fish-Culturist* 34: 240–241.

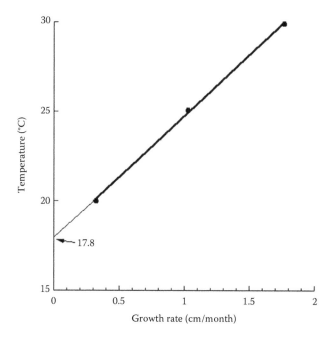

Figure 2.3 Growth of blue tilapia in centimeters per month at temperatures from 20 to 30°C. (From Soderberg, R. W. 1990. *Progressive Fish-Culturist* 52: 155–157. With permission.)

An average of 6.3 MTU were required per cm of catfish growth at temperatures from 24 to 30°C.

Soderberg (1990) found that the growth rate versus temperature plot for blue tilapia (*Oreochromis aureas*) had an intercept of 17.8°C (Figure 2.3).

An average of 6.9 centigrade TU were required per cm of blue tilapia growth in temperatures of 20–30°C when temperature units available per month equal the average monthly water temperature minus 17.8°C (Table 2.7).

Table 2.7 TU (1 TU = 1°C greater than 17.8°C for 1 month) required per centimeter of blue tilapia growth at temperatures from 20 to 30°C

Exposure temperature	MTU required/cm of growth
20	6.7
25	7.0
30	6.9
Mean	6.9

Source: Data from Soderberg, R. W. 1990. *Progressive Fish-Culturist* 52: 155–157.

Table 2.8 MTU (average monthly temperature, °C) required per centimeter of tiger muskellunge early and advanced fingerling growth at temperatures from 14 to 28°C

Exposure temperature	MTU required/cm of growth	
	Early fingerlings (3–4 cm)	Advanced fingerlings (12–13 cm)
14	3.9	–
16	3.8	–
18	3.8	5.0
19	3.7	–
20	3.7	4.8
21	3.9	5.0
22	3.7	5.2
23	–	5.1
24	3.8	5.3
26	–	6.7
28	–	9.3
Mean	3.8	5.8

Source: After Meade, J. W., W. F. Krise, and T. Ort. 1983. *Aquaculture* 32: 157–164.

Meade et al. (1983) studied the temperature-related growth of tiger muskellunge (*Esox lucius* × *E. masquinongy*). In the temperature range of 14–24°C, 3–4 cm early fingerlings required 3.8 MTU/cm of growth. Larger fish (12–13 cm) required 5.8 MTU/cm from 18 to 28°C and 5.1 MTU/cm from 18 to 24°C (Table 2.8).

The temperature unit and monthly temperature unit requirements of cultured fish are summarized in Table 2.9.

Regression models for fish growth

Because of the confusion associated with multiple temperature unit models with different intercepts and units of measure, Soderberg (1990) proposed that the independent variable, temperature, be regressed against growth in accordance with the general linear model. The increase in length value (ΔL) required for fish growth projection can then be calculated from the regression equation $Y = a + bX$, where $Y = \Delta L$, a = the intercept, b = the slope, and X = the fixed variable, temperature. The reversed plot of Soderberg's (1990) tilapia data is shown in Figure 2.4.

Soderberg (1992) subjected the available fish growth data to regression analysis in order to develop linear growth models for brook trout, rainbow trout, lake trout, steelhead trout, Atlantic salmon, channel catfish, tiger muskellunge, blue tilapia, and Nile tilapia.

Table 2.9 Summary of temperature unit (TU) or monthly temperature unit (MTU) requirements for cultured fish

Fish species	Temperature range (°C)	TU or MTU/cm	Reference
Brook trout	6–11	16.2[a]	Haskell (1959)
Brook trout	7–19	7.2[b]	Dwyer et al. (1981a)
Brook trout	10–16	6.1[b]	Dwyer et al. (1981a)
Rainbow trout	7–19	7.1[b]	Dwyer et al. (1981b)
Lake trout	4–16	5.4[b]	Dwyer et al. (1981c)
Steelhead trout	4–19	10.9[b]	Dwyer et al. (1982)
Steelhead trout	13	8.5[b]	Dwyer et al. (1982)
Atlantic salmon	7–16	10.1[b]	Dwyer and Piper (1987)
Channel catfish	20–30	6.3[b]	Andrews et al. (1972)
Blue tilapia	20–30	6.9[c]	Soderberg (1990)
Tiger muskellunge			
3–4 cm	14–24	3.8[b]	Meade et al. (1983)
12–13 cm	18–28	5.8[b]	Meade et al. (1983)
12–13 cm	18–24	5.1[b]	Meade et al. (1983)

[a] TU = Mean monthly temperature in °F–38.6°F.
[b] MTU = Mean monthly temperature in °C.
[c] TU = Mean monthly temperature in °C–17.8°C.

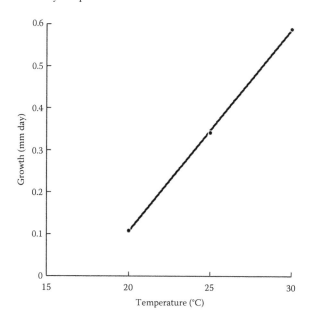

Figure 2.4 Linear growth plot for blue tilapia. $\Delta L = -0.853 + 0.048\ T$ ($r^2 > 0.99$). (Data from Soderberg, R. W. 1990. *Progressive Fish-Culturist* 52: 155–157.)

Brook trout: Growth at 4°C was significantly less efficient than at temperatures from 7 to 19°C (Dwyer et al. 1981a). The growth equation for the temperature range of 4–19°C is $\Delta L = 0.155 + 0.0355$ T (ΔL is in millimeters (mm)/day and T is in °C for all equations presented here) with a coefficient of determination (r^2) of 0.637. The r^2 value is improved to 0.792 in the more efficient temperature range of 7–19°C. The growth equation over this temperature range is $\Delta L = 0.006 + 0.0455$ T. Although temperature unit requirements did not differ significantly over the temperature range of 7–19°C, actual growth was less at 19°C than at 16°C (Dwyer et al. 1981a). Thus, the linear growth model is further improved by narrowing the temperature range to 7–16°C, where $\Delta L = -0.068 + 0.0578$ T ($r^2 = 0.882$). The values derived from Haskell's (1959) data are better correlated ($r^2 = 0.99$) than those of Dwyer et al. (1981a), probably due to the narrower temperature range (5.5–11.1°C) of his determinations. The equation resulting from regression analysis of Haskell's data is $\Delta L = 0.348 + 0.0944$ T and gives ΔL values similar to those predicted by the equation derived from Dwyer's data in the temperature range of 7–16°C.

Rainbow trout: The MTU requirements were significantly greater at 4°C than the other test temperatures up to 19°C. Growth efficiency did not differ significantly in the temperature range of 7–19°C, but actual growth at 19°C was less than that at 16°C (Dwyer et al. 1981b). The growth equations for the temperature ranges of 4–19°C, 7–19°C, and 7–16°C are $\Delta L = -0.040 + 0.505$ T ($r^2 = 0.886$), $\Delta L = 0.043 + 0.0450$ T ($r^2 = 0.801$), and $\Delta L = -0.167 + 0.066$ T ($r^2 = 0.971$), respectively.

Lake trout: The MTU requirements for growth did not significantly differ over the temperature range of 4–16°C (Dwyer et al. 1981c), but fish at 16°C grew less than those at 13°C. The linear growth equation for the temperature range of 4–16°C is $\Delta L = 0.176 + 0.0426$ T ($r^2 = 0.858$), and the equation for the temperature range of 4–13°C is $\Delta L = 0.0622 + 0.588$ T ($r^2 = 0.979$).

Steelhead trout: The MTU requirement per centimeter of growth was significantly lower at 13°C than at 4, 7, 10, 16, or 19°C (Dwyer et al. 1982). The regression equation over the temperature range of 4–19°C is $\Delta L = 0.0329 + 0.0294$ T ($r^2 = 0.856$). The growth equation is $\Delta L = 0.0148 + 0.0343$ T ($r^2 = 0.937$) over the temperature range of 7–16°C.

Atlantic salmon: Fish grown at 4 and 19°C required significantly more MTU per unit of length increase than those held at 7, 10, 13, and 16°C. Growth at 19°C was less than that at 16°C (Dwyer and Piper 1987). The growth equation over the entire range of temperatures tested is $\Delta L = 0.0043 + 0.0306$ T ($r^2 = 0.926$). The improved regression over the temperature range of 7–16°C is $\Delta L = -0.0429 + 0.0371$ T ($r^2 > 0.99$).

Channel catfish: Andrews et al. (1972) studied growth at temperatures from 24 to 30°C. Calculated daily incremental growth increased with temperature from 24 to 28°C, but was less at 30 than at 28°C. The regression equation for catfish growth is $\Delta L = 0.612 + 0.0298$ ($r^2 = 0.825$) at temperatures from 24 to 30°C and $\Delta L = 0.195 + 0.0463$ T ($r^2 = 0.991$) from 24 to 28°C.

Tiger muskellunge: The growth of early fingerlings (3–4 cm) was well correlated with temperature in the range of 14–24°C (Meade et al. 1983). The regression equation calculated from their data is $\Delta L = -0.0548 + 0.912\ T$ ($r^2 = 0.985$). The growth rate of advanced fingerlings (12–13 cm) was greatest at 23 and 24°C but declined at higher and lower temperatures (Meade et al. 1983). Thus, temperature and daily incremental growth for larger tiger muskellunge were not correlated ($r^2 = 0.12$) over the entire range of 18–28°C. The growth equation for the temperature range 18–24°C is $\Delta L = 0.394 + 0.0471$ ($r^2 = 0.864$).

Blue tilapia: Soderberg (1990) reported that the regression of daily growth increment on temperature in the range of 20–30°C is $\Delta L = -0.853 + 0.048\ T$ ($r^2 > 0.99$).

Nile tilapia (*Oreochromis niloticus*): Soderberg (2006) grew Nile tilapia of two sizes at temperatures from 21 to 30°C in two different years. The resulting regression lines were not significantly different (P > 0.05), so he combined the data from both years to report that the regression of daily growth increment on temperature in the range of 21–30°C is $\Delta L = -1.6707 + 0.0968\ T$ ($r^2 = 0.95$) (Figure 2.5).

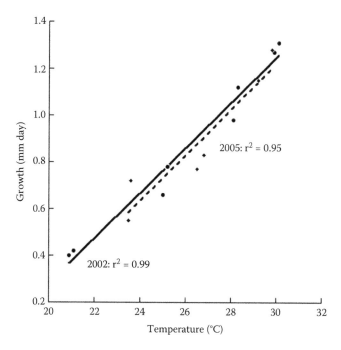

Figure 2.5 Regression of daily growth rate against temperature for Nile tilapia in two different years. The slopes of the regression lines are not significantly different (P > 0.05). The combined equation is $\Delta L = -1.6707 + 0.0968$ ($r^2 = 0.95$). (Data from Soderberg, R. W. 2006. *North American Journal of Aquaculture* 68: 245–248.)

Table 2.10 Linear models for fish growth in hatcheries

Species	Temperature range (°C)	Growth model	r^2	Reference
Brook trout	5.5–11.1	$\Delta L = -0.348 + 0.094\ T$	>0.99	Haskell (1959)
Brook trout	4–19	$\Delta L = 0.155 + 0.035\ T$	0.637	Dwyer et al. (1981a)
Brook trout	7–19	$\Delta L = 0.006 + 0.045\ T$	0.792	Dwyer et al. (1981a)
Brook trout	7–16	$\Delta L = -0.068 + 0.058\ T$	0.882	Dwyer et al. (1981a)
Rainbow trout	4–19	$\Delta L = -0.040 + 0.051\ T$	0.886	Dwyer et al. (1981b)
Rainbow trout	7–19	$\Delta L = 0.043 + 0.045\ T$	0.801	Dwyer et al. (1981b)
Rainbow trout	7–16	$\Delta L = -0.167 + 0.066\ T$	0.971	Dwyer et al. (1981b)
Lake trout	4–16	$\Delta L = 0.176 + 0.043\ T$	0.858	Dwyer et al. (1981c)
Lake trout	4–13	$\Delta L = 0.062 + 0.059\ T$	0.979	Dwyer et al. (1981c)
Steelhead trout	4–19	$\Delta L = 0.033 + 0.029\ T$	0.856	Dwyer et al. (1982)
Steelhead trout	7–16	$\Delta L = 0.015 + 0.034\ T$	0.937	Dwyer et al. (1982)
Atlantic salmon	4–19	$\Delta L = 0.004 + 0.031\ T$	0.926	Dwyer and Piper (1987)
Atlantic salmon	7–16	$\Delta L = -0.043 + 0.037\ T$	0.999	Dwyer and Piper (1987)
Channel catfish	24–30	$\Delta L = 0.612 + 0.030\ T$	0.825	Andrews et al. (1972)
Channel catfish	24–28	$\Delta L = 0.195 + 0.046\ T$	0.991	Andrews et al. (1972)
Tiger muskellunge				
3–4 cm	14–24	$\Delta L = -0.055 + 0.091\ T$	0.985	Meade et al. (1983)
12–13 cm	18–24	$\Delta L = -0.039 + 0.047\ T$	0.864	Meade et al. (1983)
Blue tilapia	20–30	$\Delta L = -0.853 + 0.048\ T$	>0.99	Soderberg (1990)
Nile tilapia	21–30	$\Delta L = -0.671 + 0.097\ T$	0.950	Soderberg (2006)

Source: Data from Soderberg, R. W. 1992. *Progressive Fish-Culturist* 54: 255–258; Soderberg, R. W. 2006. *North American Journal of Aquaculture* 68: 245–248.

Note: T = temperature in °C and ΔL = daily growth in mm. r^2 is the coefficient of determination.

The regression equations described here and their coefficients of determination are summarized in Table 2.10.

Projection of fish growth in time

Temperature unit procedure

The ability to project the size of fish in advance is necessary for determining daily feed rates, feed orders, egg orders, stocking dates, and production schedules. The following example illustrates how temperature unit methods are used to predict fish growth. Consider a lot of brook trout being reared at a hatchery where the average monthly water temperature is 8.2°C. If 7.2 MTU are required per centimeter of growth, the expected monthly gain in length is

$$\frac{8.2\,\text{MTU}}{\text{month}} \times \frac{\text{cm}}{7.2\,\text{MTU}} = \frac{1.14\,\text{cm}}{\text{month}}$$

Similarly, using TU, if the average monthly water temperature is 49°F, 49 − 38.6, or 10.4 TU are available for that month. If 17 TU are required per inch of growth, the monthly length gain is

$$\frac{10.4\,\text{TU}}{\text{month}} \times \frac{\text{inch}}{17\,\text{TU}} = \frac{0.61\,\text{inch}}{\text{month}} \times \frac{2.54\,\text{cm}}{\text{inch}} = \frac{1.55\,\text{cm}}{\text{month}}$$

Fish growth is projected in units of length, but the weight of fish of known length is often required for fish hatchery management. Weight is easily calculated from length with the following expression:

$$W = KL^3 \qquad\qquad (2.1)$$

where W = fish weight, L = fish length, and K = condition factor. The condition factor is species- and strain-dependent but remains constant through the linear phase of growth if diet and health are adequate.

Haskell (1959) determined the value of K to be 0.0004 for brook, brown, and rainbow trout when W is given in pounds and L in inches. Piper et al. (1982) provided English-unit K values for some other fish species. These are summarized in Table 2.11.

Existing K values converted to International System of Units (SIU), along with those given by Soderberg (1990) for blue tilapia and Soderberg (2006) for Nile tilapia, are given in Table 2.12.

An example of fish weight projection follows: Suppose a hatchery manager wishes to estimate the monthly gain of 20,000 12-cm channel catfish being reared at 28°C. The MTU required per centimeter of catfish length gain is 6.3 (Table 2.6). Thus, the expected monthly growth is $28/6.3 = 4.4$ cm, and the fish should be 16.4 cm long at the end of the month. The weight of a 12-cm catfish is estimated to be $W = 0.00798 \times 12^3 = 13.8$ g (Table 2.12), and the weight of a 16.4-cm catfish is estimated to be $W = 0.00798 \times 16.4^3 = 35.2$ g. The expected weight gain is $35.2 − 13.8 = 21.4$ g, and the expected gain of 20,000 fish is 428 kg.

Regression model procedure

Regression equations provide a less confusing method for the calculation of ΔL than temperature unit procedures. They should also be

Table 2.11 Average K values for some species of cultured fish

Species	K × 10⁴	Reference
Brook, brown, and rainbow trout	4.055	Haskell (1959)
Muskellunge (*Esox masquinongy*)	1.600	Piper et al. (1982)
Northern pike (*E. lucius*)	1.811	Piper et al. (1982)
Lake trout	2.723	Piper et al. (1982)
Chinook salmon (*Oncorhynchus tshawytscha*)	2.959	Piper et al. (1982)
Walleye (*Sander vitreus*)	3.000	Piper et al. (1982)
Channel catfish	2.877	Piper et al. (1982)
Cutthroat trout (*O. clarkii*)	3.559	Piper et al. (1982)
Coho salmon (*O. kisutch*)	3.737	Piper et al. (1982)
Steelhead trout	3.405	Piper et al. (1982)
Largemouth bass (*Micropterus salmoides*)	4.606	Piper et al. (1982)
Blue tilapia	8.430	Soderberg (1990)
Nile tilapia	6.696	Soderberg (2006)

Note: $K = WL^3$, where W = weight in pounds and L = length in inches.

more accurate than temperature unit models that do not correct for the zero-growth intercept. The first step in the use of these equations is to select the one with the highest coefficient of determination whose temperature range includes the temperature for which ΔL is required. For example, if the growth of rainbow trout at 10°C is required, the equation $\Delta L = -0167 + 0.666 \ T$ ($r^2 = 0971$) (Table 2.10) should be used because the

Table 2.12 Average K values for some species of cultured fish

Species	K × 10⁴	Reference
Brook, brown, and rainbow trout	0.01125	Haskell (1959)
Muskellunge	0.00444	Piper et al. (1982)
Northern pike	0.00503	Piper et al. (1982)
Lake trout	0.00756	Piper et al. (1982)
Chinook salmon	0.00821	Piper et al. (1982)
Walleye	0.00833	Piper et al. (1982)
Channel catfish	0.00798	Piper et al. (1982)
Cutthroat trout	0.00988	Piper et al. (1982)
Coho salmon	0.01037	Piper et al. (1982)
Steelhead trout	0.00945	Piper et al. (1982)
Largemouth bass	0.01278	Piper et al. (1982)
Blue tilapia	0.02330	Soderberg (1990)
Nile tilapia	0.0186	Soderberg (2006)

Note: $K = WL^3$, where W = weight in grams and L = length in centimeters.

other two equations for this species were determined over greater temperature ranges and thus have lower coefficients of determination. Once the ΔL value is calculated, it is used in the projection of fish growth as previously demonstrated.

When fish are inventoried, the projected length is compared to the actual length and the ΔL value is adjusted to account for any hatchery-specific deviations from the growth model.

Feeding intensively cultured fish

The nutritional requirements of fish are not completely known, but adequate commercial diets are available for trout, salmon, and catfish. These species are carnivorous and require high-protein diets because they digest carbohydrates rather poorly compared to other livestock. Available calories (C) per gram (g) of protein, fat, and carbohydrates for trout are approximately 3.9, 8.0, and 1.6, respectively (Piper et al. 1982). Approximately 3850 calories are required per kilogram (kg) of trout weight gain (Piper et al. 1982). Thus, the food conversion, or the amount of food required per unit of fish weight gain, can easily be estimated if the proximate analysis of the diet is known. For example, suppose a trout diet contains 45% protein, 8% fat, and 10% carbohydrate. The available calories in a kg of this diet are:

Protein	$450 \text{ g/kg} \times 3.9 \text{ C/g}$	$=$	1755 C/kg
Fat	$80 \text{ g/kg} \times 8.0 \text{ C/g}$	$=$	640 C/kg
Carbohydrate	$100 \text{ g/kg} \times 1.6 \text{ C/g}$	$=$	160 C/kg
Total			2555 C/kg

The anticipate food conversion then is

$$\frac{3850 \text{ C per kg fish}}{2555 \text{ C per kg food}} = \frac{1.51 \text{ kg food}}{\text{kg gain}}$$

Modern trout and salmon diets typically contain 42%–45% protein and 16%–20% fat. These formulations result in food conversion values of around 1.0.

Channel catfish, and probably other warmwater fish, digest components of their diet differently than trout. Available calories per gram of protein, fat, and carbohydrate are 4.5, 8.5, and 2.9, respectively. Approximately 2156 calories are required per kilogram of catfish gain (Piper et al. 1982).

The food conversion is most commonly determined for a particular diet and fish strain from past station records. Calculation of the estimated food conversion is useful for evaluating potential diets, beginning a feeding schedule for new hatcheries, or investigating poor feeding efficiency.

Haskell (1959) introduced a feeding equation

$$F = \frac{3 \times C \times \Delta L \times 100}{L} \qquad (2.2)$$

where F = percent body weight to feed per day, C = food conversion, ΔL = daily increase in length, and L = length of fish on day fed. Notice that the feed rate equation calculates the amount of food required to result in a particular increment of growth. Normally, the ΔL value will be the maximum potential growth rate of a particular fish species at a certain growth temperature. If production schedules call for slower growth rates to result in a particular size of fish on a given date, a smaller ΔL value may be substituted.

Butterbaugh and Willoughby (1967) noticed that when the diet, fish strain, and water temperatures were unchanged, the factors in the numerator of the feed equation remained constant. They introduced a hatchery constant, HC, which combines these factors in a single term. Thus,

$$HC = 3 \times C \times \Delta L \times 100 \qquad (2.3)$$

and

$$F = \frac{HC}{L} \qquad (2.4)$$

Hatchery records

Daily food rations for each lot of fish are calculated, and growth is projected each day by adding ΔL to L. Usually a sample count is taken approximately every 3 months when fish are graded. At this time, the food conversion can be calculated and the projected fish size can be corrected with the actual value obtained in the sample count.

SAMPLE PROBLEMS

1. You know from past station records that 1.0 kg of food is required for each kilogram of fish weight gain. Your facility holds 20,000 25-cm rainbow trout on June 1, with a K factor of 0.01125. You are assigned to write up the feed orders for July, August, and September. The hatchery manager tells you that the expected water temperatures for these months are 12, 13, and 11°C, respectively. How much food should you order?

2. A hatchery manager is requested to supply 30-cm fish on April 1 for the opening day of trout season. The water temperature at his hatchery is a constant 12°C. When should he have eggs delivered, assuming that they hatch as soon as they arrive, and the fry are 1.5 cm long at that time? Assume a 2-week swim-up time. The swim-up time is the period between when the eggs hatch and when the fry first take feed.

3. Calculate the cost of producing a 25-cm brook trout given the following information:

   Length at first feeding is 2.0 cm
   K = 0.01125
   Food conversion is 1.3
   Food cost is $430/tonne (metric ton)

4. Tiger muskies, the cross between a muskellunge and northern pike, are now grown routinely in hatcheries on dry feed. Newly hatched fry are 0.7 cm in length and begin feeding immediately upon hatching; they can reach a length of 20 cm in the 3-month summer season. What is the hatchery constant for a hatchery experiencing this growth rate and a food conversion of 1.7?

5. For the example above, how much food is required to raise 100,000 20-cm tiger muskies?

6. You are assigned to feed fish at a brown trout hatchery. Given the following information, how much feed do you require on August 15?

   Lot #1: Sample count on August 1: 946 fish weighed 5.2 kg. Total number of fish in lot is 22,501.

   Lot #2: Sample count on August 1: 501 fish weighed 23.4 kg. Total number of fish in lot is 19,055.

   Lot #3: Sample count on August 1: 200 fish weighed 98.1 kg. Total number of fish in lot is 51,250.

   Hatchery constant is 8.55. Last year, this hatchery produced 150,000 kg of fish on 198,000 kg of food.

7. For the following lot of lake trout, calculate:
   a. The hatchery constant
   b. The food ration for August 1
   c. The food ration for August 30

   Estimated food conversion is 1.1, water temperature is 11.5°C, and sample count on July 15 is as follows: 936 fish weighed 158 kg. The lot contains 215,000 fish.

8. Calculate the food requirement for the following lot of catfish being grown in a geothermal spring.

   Sample count: 1026 fish weighed 251 kg
   Water temperature is 28°C
   C = 1.7
   Lot contains 110,000 fish

9. Choose the most economical diet given the following information:
   Diet A costs \$370/tonne and gives a food conversion of 1.4.
   Diet B costs \$430/tonne and gives a food conversion of 1.2.
   Diet C costs \$500/tonne and gives a food conversion of 1.0.
10. Choose the most economical trout diet from those listed below.

Diet	% Protein	% Fat	% Carbohydrate	Cost (\$/kg)
A	38	5	16	0.45
B	42	12	8	0.48
C	45	16	5	0.51
D	48	20	4	0.62

References

Andrews, J. W., L. H. Knight, and T. Murai. 1972. Temperature requirements for high density rearing of channel catfish from fingerlings to market size. *Progressive Fish-Culturist* 34: 240–241.

Butterbaugh, G. L. and H. Willoughby. 1967. A feeding guide for brook, brown and rainbow trout. *Progressive Fish-Culturist* 29: 210–215.

Dwyer, W. P. and R. G. Piper. 1987. Atlantic salmon growth efficiency as affected by temperature. *Progressive Fish-Culturist* 49: 57–59.

Dwyer, W. P., R. G. Piper, and C. E. Smith. 1981a. *Brook Trout Growth Efficiency as Affected by Temperature.* Information Leaflet 16, U.S. Fish and Wildlife Service Fish Culture Development Center, Bozeman, Montana.

Dwyer, W. P., R. G. Piper, and C. E. Smith. 1981b. *Rainbow Trout Growth Efficiency as Affected by Temperature.* Information Leaflet 18, U.S. Fish and Wildlife Service Fish Culture Development Center, Bozeman, Montana.

Dwyer, W. P., R. G. Piper, and C. E. Smith. 1981c. *Lake Trout, Salvelinus namaycush, Growth Efficiency as Affected by Temperature.* Information Leaflet 22, U.S. Fish and Wildlife Service Fish Culture Development Center, Bozeman, Montana.

Dwyer, W. P., R. G. Piper, and C. E. Smith. 1982. *Steelhead Trout Growth Efficiency as Affected by Temperature.* Information Leaflet 27, U.S. Fish and Wildlife Service Fish Culture Development Center, Bozeman, Montana.

Haskell, D. C. 1959. Trout growth in hatcheries. *New York Fish and Game Journal* 6: 205–237.

Meade, J. W., W. F. Krise, and T. Ort. 1983. Effect of temperature on the growth of tiger muskellunge in intensive culture. *Aquaculture* 32: 157–164.

Piper, R. G., J. B. McElwain, L. E. Orme, J. P. McCraren, L. G. Fowler, and J. R. Leonard. 1982. *Fish Hatchery Management.* U.S. Fish and Wildlife Service, Washington, DC.

Soderberg, R. W. 1990. Temperature effects on the growth of blue tilapia in intensive culture. *Progressive Fish-Culturist* 52: 155–157.

Soderberg, R. W. 1992. Linear fish growth models for intensive aquaculture. *Progressive Fish-Culturist* 54: 255–258.

Soderberg, R. W. 2006. A linear fish growth model for Nile tilapia in intensive aquaculture. *North American Journal of Aquaculture* 68: 245–248.

chapter three

Water sources for flowing water fish culture

The hydrologic cycle

> All the rivers run into the sea, yet the sea is not full;
> unto the place from whence the rivers come, thither
> they return again

Ecclesiastes 1:7

Mankind has long been aware of the hydrologic cycle—the processes of evaporation, transportation, condensation, precipitation, percolation, and runoff—by which water is recycled through the hydrosphere.

Evaporation occurs when radiant heat is absorbed and converted to kinetic energy in water molecules. The resulting molecular velocity causes some molecules to leave the water surface and enter the atmosphere. The energy lost in this process is the latent heat of vaporization, and the water surface cools. In addition to solar radiation, the rate of evaporation is affected by humidity, which determines the vapor pressure gradient; water turbidity, which may affect the amount of heat absorbed at the water surface; and, most importantly, by wind, which carries away the saturated layer of air at the water's surface, maintaining the vapor pressure gradient. Plants increase the surface area over which evaporation occurs. The evapotranspiration rate from aquatic plants may be twice as high as evaporation from the free water surface.

Condensation is caused by temperature changes in the air that affect its ability to hold moisture. The dew point is the temperature at which air becomes saturated with water. As air rises, it expands and cools according to the fundamental gas laws (see Chapter 5). When the dew point is reached, condensation occurs and clouds form. The clouds are made up of small droplets of liquid water too small to fall as rain. Precipitation occurs when ice crystals, dust, or sea salts act as hydroscopic nuclei upon which water droplets coalesce into raindrops. Air rises, causing precipitation by convective warming, by moving wedges of cold air that cause frontal storms, or by orographic processes in which air rises over geographic features.

Some rainfall is intercepted by plants and other obstacles and may evaporate before reaching the soil surface. The remainder enters the soil by percolation or runs off the soil surface into streams. The percolation rate is affected by gravity, capillarity, soil moisture content, and the amount of air in the soil pores. The layer of soil that retains soil moisture is called the zone of rock fracture. When this zone is saturated with water, infiltration continues to the zone of rock storage, where it is referred to as groundwater.

Stream flow consists of runoff, channel precipitation, and base flow, minus water losses to consumption, evaporation, and seepage. Base flow is the contribution to stream discharge from groundwater. The amount of precipitation that runs off a particular watershed is referred to as the hydrologic response and is expressed as a percentage. The hydrologic response is calculated from a hydrograph, which measures direct runoff above base flow following a precipitation event and from rain gauges in the watershed, which measure total precipitation. Hydrologic response values may range from 1% to 75% and are related to soil permeability, bedrock porosity, slope, and vegetation. Mountainous areas with shallow soils and impermeable bedrock have high hydrologic responses because a relatively large fraction of the water that falls as precipitation enters streams. Low hydrologic responses occur in areas where most of the precipitation enters the groundwater due to low gradients, permeable soils, and porous bedrock, such as limestone or shale.

Over 97% of the water on Earth is in the oceans and over 77% of the remaining freshwater is locked in ice caps and glaciers. Some intensive aquacultures involve the pumping of seawater to rearing units on land, but the water supplies usually used for flowing water fish culture are relatively shallow groundwaters and surface waters, usually from streams. These sources comprise a relatively small fraction of the water circulating through the hydrologic cycle and, in the case of streams, a very transient one.

Use of surface waters for fish culture

In areas of high hydrologic response, streams may be diverted through rearing units for flowing water fish culture. Stream discharge is composed of runoff and base flow, but in areas where a significant base flow occurs in streams, groundwater is often chosen as the aquaculture water supply. Thus, the use of surface water is most practical in locations that receive regular and heavy rainfall and have high hydrological responses.

Water temperatures of streams with low base flows vary seasonally with the air temperature and the degree of variation depending primarily upon the magnitude of base flow component and summer shading by riparian vegetation. Thus, a moderate climate and wooded watersheds are required for coldwater fish culture using stream water supplies and a tropical climate is necessary for similar warmwater fish culture. Streams are used as water supplies for trout culture in the southern Appalachian region of the United States. Suitable locations are higher than 1000 m above sea level because below this elevation, summer water temperatures may get too warm for trout survival and the winter period of reduced trout growth is excessive. Even at higher elevations, the growing season is limited to about 8 months due to high summer and low winter temperatures (Figure 3.1).

A few experimental catfish farms in the southern United States have used reservoir water in flowing water designs that is then pumped back to the reservoir after passage through the rearing units. This technology should be economically comparable to traditional static water catfish culture. Stream waters have not been used for intensive warmwater fish culture in temperate climates because they achieve temperatures suitable for satisfactory fish growth only for brief summer periods. Use of surface water for flowing water fish culture is appropriate for low-elevation tropical locations due to constant warm temperatures. A significant industry occurs in Central America using surface waters to produce tilapia (Figure 3.2). Fish are exported to Western markets, which makes the use of the required artificial diets profitable.

Environmental regulations in some areas may prohibit stream diversion, thereby precluding their use for fish culture.

Figure 3.1 Intake structure diverting surface water to a trout farm in North Carolina.

Figure 3.2 Flowing water tilapia farm in Costa Rica supplied with surface water. (Photo courtesy of Aaron McNevin.)

Use of groundwater for fish culture

Groundwater is water stored in geologic material. The zone of saturation in water-bearing strata is called an aquifer. Aquifers are characterized by their porosity, which is related to the void fraction of the stratum, and their permeability, which describes the degree of water movement through the aquifer and is related to pore size. Water-bearing strata may be well-sorted, such as sand, or poorly sorted aggregates of different-sized particles. In some aquifers, mineral deposits may partially fill the voids. In rock aquifers, water may be held in channels where the rock has dissolved or in voids created by rock fracture. The amount of water in saturated strata may range from 0.01% to 0.35% by volume.

Groundwater is usually chosen for flowing water fish culture because it has a nearly constant temperature throughout the year. This is because the temperature of the aquifer is maintained at approximately the mean temperature of the water entering it. It follows that the temperature of the groundwater can be estimated from the annual mean air tempera-ture of a region. This generalization holds true as long as groundwaters are not influenced by geothermal activity. Warmwater fish are occasion-ally grown in temperate climates using geothermal wells or springs or industrially heated wastewater (Figure 3.3).

In temperate climates, groundwaters are much warmer than surface waters in the winter and much cooler than surface waters in the summer. For this reason, groundwater is our most usable source of solar energy. Average groundwater temperatures in the United States are shown in

Figure 3.3 Channel catfish farm in Idaho using geothermally heated water from an artesian well.

Figure 3.4. Note that the isotherms follow latitudinal as well as elevational determinants of mean air temperature. Thus, equatorial areas at high elevations, such as those found in South America and Africa, often have suitable groundwater as well as surface water for coldwater fish culture (Figure 3.5).

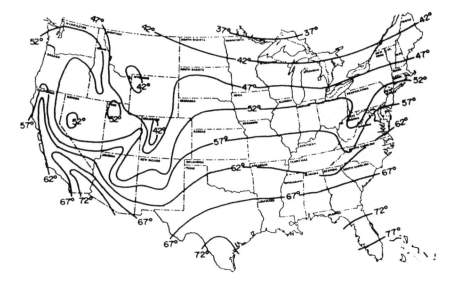

Figure 3.4 Groundwater isotherms for the United States.

Figure 3.5 Tilapia farm in Kenya supplied with pumped groundwater. The water is a suitable temperature for warmwater fish due to the equatorial, coastal location of the site.

Aquifers are recharged by precipitation that percolates through the soil to the water-bearing stratum or, in some cases, by streams that flow into permeable formations. Groundwater may be accessed at springs where it erupts through the surface or by wells driven into the aquifer. If a well is drilled into an unconfined aquifer (one that is not under pressure), water will fill the pipe to the upper surface of the aquifer or water table and must be pumped. In a confined aquifer, the water is under pressure caused by the head differential from the recharge area to the well. If this aquifer is accessed by a well, water will rise in the pipe to the piezometric surface, which is a level determined by the amount of head differential and the frictional head losses that occur as the water moves through the aquifer. Such wells are called artesian and may be desirable for aquaculture water supplies because the pumping depth may be less than that from wells in unconfined aquifers. If the piezometric surface is above ground level, water will flow from an artesian well without pumping. Discharges from flowing wells are sometimes increased by pumping.

When a streambed intersects an aquifer, a considerable portion of its discharge may be base flow. If the stream bottom lies above the water table, water seeps out of the stream toward the aquifer. The resulting elevated mound of groundwater is called bank storage and may extend for several miles from the beds of large rivers.

When a well is pumped, the area around it becomes dewatered and the water level drops. When equilibrium between discharge of the well and recharge of the pumped area from the aquifer is reached there will be a cone-shaped dewatered area around the well. This area is called the cone of depression, and its bottom is the actual depth from which water must be pumped. The dimensions of the cone of depression are determined by the permeability of the aquifer and the recharge rate. Wells drilled into highly permeable strata will have wide, shallow cones of depression while those in low permeability strata will have narrow, deep cones of depression, resulting in considerable drawdown during pumping and consequent increases in pumping depth. Because pumping cost is related to the depth from which water is pumped, the shape of the cone of depression is an important consideration in the economic evaluation of a well for aquaculture. A test well must be drilled and operated in order to determine the actual pumping depth from a particular aquifer. In areas where wells are closely spaced, cones of depression may overlap, causing localized lowering of the water table.

The pump size required for a particular job is calculated from the conversion factor of 33,000 foot-pound (ft lb)/minute (min) of work per horsepower (HP). Suppose we wish to pump 1000 gallons per minute (gpm) from 100 feet. The horsepower requirement is

$$\frac{1000 \text{ gpm} \times (8.34 \text{ lb/gallon}) \times 100 \text{ ft}}{(33,000 \text{ ft lb/min})/\text{HP}} = 25 \text{ HP}$$

Determination of required pump power in the metric system involves converting the mass in kg to Newtons by multiplying by the acceleration due to gravity, 9.8 m/sec². The required pump power in kilowatts (kW) to lift 0.063 m³/sec 30.5 m is

$$\frac{0.063 \text{ m}^3}{\text{sec}} \times \frac{1000 \text{ kg}}{\text{m}^3} \times 30 \text{ m} \times \frac{9.8 \text{ m}}{\text{sec}^2} \times \frac{1 \text{ Joule (J)}}{\text{kg m}^2/\text{sec}^2} \times \frac{\text{kW}}{1000 \text{ J/sec}} = 18.5 \text{ kW}$$

The efficiency of the pump engine describes the relationship between water HP, which we have just calculated, and brake HP, which the pump engine must develop. If the motor is 75% efficient, typical of electrically powered units, $25/0.75 = 33$ HP ($18.65/0.75 = 24.9$ kW) is the amount of power that the pump requires. Pumping cost in English units is calculated by converting HP to kilowatts. Electricity is purchased by the kilowatt hour (kWh). For the present example, using electrical power purchased for \$0.14/kWh, the annual pumping cost is

$$33 \text{ HP} \times \frac{0.746 \text{ kW}}{\text{HP}} \times \frac{8640 \text{ h}}{\text{year}} \times \frac{\$0.14}{\text{kWh}} = \$29,778/\text{year}$$

The pumping cost for a diesel pump is determined as follows, assuming that the engine is 50% efficient and the fuel contains 138,700 British Thermal Units (BTU) per gallon and is purchased for $2.00/gallon. The conversion from BTU to HP is 2544.

$$\frac{25\,\text{HP}}{0.50} \times \frac{2544\,\text{BTU}}{\text{HP h}} \times \frac{\text{gallon}}{138,700\,\text{BTU}} \times \frac{8640\,\text{h}}{\text{year}} \times \frac{\$2.00}{\text{gallon}} = \$15,847/\text{year}$$

In metric units, diesel fuel has an energy density of 10.9 kWh/L. If diesel fuel is purchased for $0.50/L, pumping cost is

$$\frac{18.5\,\text{kW}}{0.50} \times \frac{\text{L}}{10.9\,\text{kWh}} \times \frac{8640\,\text{h}}{\text{year}} \times \frac{\$0.50}{\text{L}} = \$17,467/\text{year}$$

Water flow in channels and pipes

Water flow in open channels

Water for flowing water fish culture must often be channeled or piped from its source to a suitable location for the rearing units. Flow rates through open channels are estimated by first calculating the expected velocity and then calculating the discharge from

$$Q = VA \qquad (3.1)$$

where Q = discharge in units cubed per unit of time, V = velocity in linear units over time, and A = cross-sectional area of flow in units squared.

Manning's equation is generally used to estimate water velocities in open channels:

$$V = \frac{1}{n} R^{2/3} S^{1/3} \qquad (3.2)$$

where V = velocity in meters per second (m/sec), n = roughness coefficient of the channel walls, R = hydraulic radius in m, and S = channel slope in units of fall per unit of length. The hydraulic radius is a measure of the geometric efficiency of the cross section and is equal to the cross-sectional area divided by the wetted perimeter. Thus, for a rectangular channel, $R = hb/2h + b$, where h is the height and b is the length of the base. For a trapezoidal channel, $R = hb/2h \tan(\upsilon/2)$, where υ = the angle of the channel's side. Values for n have been determined empirically and are found in Table 3.1.

Consider the flow through a winding earthen channel whose cross section most resembles a half circle with a radius of 0.2 m. Thus, the cross-sectional area is $\pi\ 0.2^2/2 = 0.063$ m^2 and the wetted perimeter is $\pi\ 0.4/2 = 0.63$ m. The head of the channel is 200 m from its end, over which the elevation difference is 0.4 m. Let us estimate how much water this channel can deliver to its downstream end. The value of n is estimated to be 0.026 (Table 3.1). R = 0.063/0.63 = 0.1. From Equation 3.2

Table 3.1 Roughness coefficients, n, for use in Manning's equation

Type of channel	n Value range
Earthen channels	
Earth, straight and uniform	0.017–0.025
Earth, winding and sluggish	0.022–0.030
Earth bottom, rubble sides	0.028–0.033
Dredged earth channels	0.025–0.035
Rock cuts, smooth and uniform	0.025–0.035
Rock cuts, jagged and irregular	0.035–0.050
Vegetated channels	
Dense, uniform, over 25 cm long	
Bermuda grass	0.040–0.200
Kudzu	0.047–0.230
Lespedeza	0.047–0.095
Dense, uniform, less than 6 cm long	
Bermuda grass	0.034–0.110
Kudzu	0.045–0.160
Lespedeza	0.023–0.050
Natural streams	
Straight banks, uniform depth	0.025–0.040
Winding, some pools, few stones or weeds	0.033–0.055
Sluggish with deep pools, few stones or weeds	0.045–0.060
Very weedy streams	0.075–0.150
Lined channels and pipes	
Concrete	
Smooth metal	0.012–0.018
Corrugated metal	0.021–0.026
Wood	0.010–0.015
Plastic	0.012–0.024

Source: Modified after Wheaton, F. W. 1977. *Aquaculture Engineering.* John Wiley & Sons, New York, New York.

$$V \doteq \frac{1}{0.026}(0.1)^{2/3}\left(\frac{0.4}{200}\right)^{1/3} = 1.0 \text{ m/sec}$$

From Equation 3.1, $Q = 1.0 \times 0.063 = 0.063$ m³/sec.

For soil- and vegetation-lined channels, erosion and sedimentation must be considered. Excessive velocities will erode the channel and low velocities could cause it to be filled in. Maximum velocities in bare-earth channels range from 0.5 to 1.5 m/sec, depending upon the soil type. Channels lined with vegetation can carry greater flows than those without vegetation.

Water flow in pipes

The conveyance method is a useful and simple procedure for determining pipe sizes and discharges for gravity-flow piped water systems, as when rearing units are fed from a head box. The conveyance of a particular pipe is calculated as

$$K = \left[6.304\left(\frac{2\log D}{e} + 1.14\right)D^{2.5}\right] \tag{3.3}$$

where $K =$ conveyance, $D =$ pipe diameter in ft, and e is a frictional coefficient related to the roughness of the pipe material. Values for e of some pipe surfaces are provided in Table 3.2. Discharge is calculated from conveyance by

$$Q = K\sqrt{\frac{h_L}{L}} \tag{3.4}$$

Table 3.2 Roughness coefficients, e, for use in conveyance equation

Pipe surface	Roughness (e)
Glass or plastic	Smooth
Iron or steel	1.5×10^{-4}
Galvanized iron	5.0×10^{-4}
Cast iron	8.5×10^{-4}
Concrete	1.1×10^{-3}–1×10^{2}

Source: Modified after Simon, A. L. 1979. *Practical Hydraulics*. John Wiley & Sons, New York, New York.

Note: For smooth pipes, e/D is approximately 1×10^{-6}, where D = pipe diameter in feet.

where Q = discharge in cubic feet per second (cfs), h_L = head loss, and L = the length of the pipe. Head loss is the elevational difference between the inlet and the outlet of the pipe. Energy is also lost by water flowing through pipeline fittings and valves. These losses are called local losses and are most conveniently expressed as equivalent pipe lengths. These values in pipe diameters are provided in Table 3.3.

Water flow in pipes may also be estimated using a modification of Manning's equation

$$Q = \frac{z}{n}D^{2.67}S^{0.5} \qquad (3.5)$$

where Q = discharge in cfs or m^3/sec, z = 0.4644 for English measurement and 0.3116 for metric, n = Manning's roughness coefficient, D = pipe diameter in ft or m, and S = slope in distance of fall per unit of length. Manning's roughness coefficients for use in Equation 3.5 are listed in Table 3.1.

An example using these procedures follows: Calculate the discharge from a 1-inch steel pipe that is 100 ft long and contains two 90° elbows and an open gate valve. The elevational difference between the inlet and the outlet (h_L) ends of the pipe is 3 ft.

1. Conveyance Method: From Equation 3.3, the conveyance of the pipe is

$$K = \left[6.304 \left(2\log \frac{0.0833}{1.5 \times 10^{-4}} + 1.14 \right) 0.0833^{2.5} \right] = 0.0837$$

The local losses in pipe diameters are 32 for each of the elbows and 17 for the valve. Thus, the total pipe length equivalency for these fittings is 0.0833 × (32 + 32 + 17) = 6.75 ft. From Equation 3.4

Table 3.3 Equivalent pipe length of local head losses in fittings and valves, expressed in number of pipe diameters

Fitting or valve	Equivalent length (Diameter)
Run Tee	20
Branch Tee	60
90° elbow, short radius	32
90° elbow, long radius	20
45° elbow	15
Open gate valve	17
Open butterfly valve	40

Source: From Hammer, M. J. 1977. *Water and Waste-Water Technology.* John Wiley & Sons, New York, New York. With permission.

$$Q = 0.0837 \sqrt{\frac{3}{106.75}} = 0.014 \text{ cfs}$$

2. Manning's Equation: From Equation 3.5

$$Q = \frac{0.4644}{0.012} \, 0.025 \, m^{2.67} \left(\frac{3}{106.75}\right)^{0.5} = 0.00034 \, m^3/\text{sec}$$

For pressurized piped systems, such as when water is supplied from a municipal source or a pressure tank, discharge can be estimated from Equation 3.1, assuming a velocity of 5 ft/sec (1.5 m/sec). For our 1-inch pipe (0.025 m diameter, 0.000507 m² cross-sectional area) in the present example, $Q = (0.000507)(1.5) = 0.00076$ m³/sec.

Water pumps

A centrifugal pump, which is appropriate for most aquaculture applications, is a cavity with an inlet and outlet for water flow in which an impeller spins to force the water in the desired direction. Pumps designed to move small volumes of water to great heights have large impellers and small cavities, while pumps designed to move large water volumes over low head pressures have small impellers turning in large cavities. Pump performance is quantified by the specific speed computed from

$$N_s = \frac{n\sqrt{Q}}{H^{0.75}} \tag{3.6}$$

where N_s = specific speed in revolutions per minute (rpm), n = the rotational speed of the shaft, Q is the discharge, and H is the total head loss. Any consistent units of measure may be used to calculate specific speed, but the values obtained are unit-specific. Pumps with low specific speeds (500 to 2000, when Q is in gpm and H is in ft; 10 to 40, when Q is in m³/sec and H is in m) are designed for small discharges and high pressures while pumps designed to deliver large volumes over low heads have high specific speeds (5000 to 15,000, when Q is in gpm and H is in ft; 100 to 300, when Q is in m³/sec and H is in m).

Suppose we wish to lift 0.1 m³/sec over a 2 m levy to supply water from a stone quarry to a trout production facility. We have an engine that runs at 1000 rpm. Ignoring head losses in the pipe and fittings

$$N_s = \frac{1000\sqrt{0.1}}{2^{0.75}} = 188$$

We require a pump with a specific speed of 188, which is within the range of low-head, high-volume pumps. Diagrams showing the particular pump design required for specific performance characteristics are available from pump manufacturers.

Pump-sizing procedures must account for head losses due to pipe friction and fittings, as previously described.

Most hydraulic engineers would consider the procedures outlined above to be an oversimplification of water supply and pump design. For a text for a course in hydraulics, this is certainly true. In actuality, the complexities of the interactions between flow, velocity, pressure, pipe, and fitting frictional losses and pump configuration are so great that pipeline, channel, and pump design become a trial and error process of calculative iteration. The procedures outlined here are meant to give the student and the practitioner of aquaculture a first-order approximation of pipe, channel, and pump sizes needed to carry required flows. General texts on practical hydrology include Anon. (1975), Simon (1979), and Todd (1964).

Measurement of water flow rate

Water flow in pipes can be measured with commercially available in-line devices. For many aquaculture applications, a stop watch and container of known volume will provide a sufficiently accurate measure of pipe discharge.

Measurement of flow through an open channel is usually accomplished by forcing the flow through a weir that has a rectangular, trapezoidal, or 90°-triangular shape. Smaller flows are best measured with a triangular weir. Care must be taken to assure that the entire water flow passes through the weir, rather than around the sides or the bottom. For a rectangular weir

$$Q = 1.84 \, Lw(h)^{1.5} \qquad (3.7)$$

where Q = discharge in m^3/sec. Lw = the width of the weir in m, and h = the height that the water rises in the weir at equilibrium. For greatest accuracy, the difference between the stream width and the width of the weir should exceed the water height in the weir by a factor of four. For a trapezoidal weir with a 4:1 slope

$$Q = 1.37 \, Lw(h)^{1.5} \qquad (3.8)$$

And a triangular weir

$$Q = 1.37 \, (h)^{2.5} \qquad (3.9)$$

SAMPLE PROBLEMS

1. A reservoir is constructed that captures all of the runoff from a 1000-hectare (Ha) watershed (1 Ha = 10,000 m²), with a hydraulic response of 10% and an average annual rainfall of 1 m. A flowing water aquaculture facility is constructed below the dam using the reservoir as a water supply. Estimate the constant water flow available in m³/sec.

2. A rainfall event results in an average of 1 cm of precipitation over an 18,000 Ha watershed. The total discharge, above base flow, from the hydrograph is 143,000 m³. What is the hydrologic response from this watershed?

3. Estimate the maximum amount of water that can be pumped from an aquifer with a recharge basin of 200 km², annual rainfall of 40 cm, and a hydrologic response of 35%. Your answer should be in m³/sec.

4. Calculate the annual cost to pump 0.2 m³/sec from a well using an electric-powered pump. The depth of the water table is 50 m and the drawdown is 12 m. The cost of electricity is 15 cents per kWh and the pump motor efficiency is 80%.

5. A 100 kW (brake) diesel-powered pump has a discharge of 0.075 m³/sec from an unconfined aquifer whose water table is 35 m deep. What is the drawdown?

6. Design a canal to carry 0.35 m³/sec of water from a spring to an aquaculture facility. A soil engineer suggests that the channel be trapezoidal in cross section with a 2:1 side slope and that the flow velocity must not exceed 0.5 m/sec.

7. Estimate the correct size of plastic pipe for a pipeline containing two 45° elbows, a run T, a 90° elbow, and an open gate valve. The length of the line is 60 m, the available head is 0.5 m, and the required discharge is 200 L/min.

8. Calculate the discharge in L/min through a rectangular concrete sluice that is 60 cm wide, 30 cm deep, and has a slope of 1:1000.

9. A catfish hatchery requires a pump to supply 200 L/min of groundwater from a well whose pumping depth is 70 m. The pump motor rotates at 1800 rpm. Calculate the specific speed and required size in kW of the pump that should be selected.

10. A stream is channeled through a triangular weir in order to measure its volume of flow. The height of water from the apex of the v-notch channel is 20 cm. What is the volume of flow?

References

Anon. 1975. *Ground Water and Wells.* Johnson Division, UOP Inc., Edward E. Johnson, Inc., St. Paul, Minnesota.

Hammer, M. J. 1977. *Water and Wastewater Technology.* John Wiley & Sons, New York, New York.

Simon, A. L. 1979. *Practical Hydraulics.* John Wiley & Sons, New York, New York.

Todd, D. K. 1964. *Ground Water Hydrology.* John Wiley & Sons, New York, New York.

Wheaton, F. W. 1977. *Aquaculture Engineering.* John Wiley & Sons, New York, New York.

chapter four

Fish culture rearing units

Flowing water fish culture is conducted in one of two general types of rearing units that differ in their hydraulic characteristics. Linear units, usually called raceways, exhibit a hydraulic pattern that approximates plug flow in which all elements of the water move with the same horizontal velocity. Circulating units, such as round tanks, have non-uniform velocities and new water is added to a mass of circulating, used water. Thus, linear units containing fish exhibit a water quality gradient from the inlet to the outlet, while circulating units have relatively uniform water quality at all positions. Linear units have uniform water velocity, while circulating units exhibit a velocity gradient as water velocities are highest at the outer rim and decrease toward the center drain.

Linear units

Haskell et al. (1960) reported unsatisfactory fish production in raceways that had throated inlets and outlets. This was due to dead spots in the pond, which resulted in poor water circulation. By injecting dye into the pond influent they showed that the water flowed along one side of the raceway, leaving a dead spot on the other side. When water entered and exited the raceway over its entire width, plug flow was achieved, and fish production improved (Haskell et al. 1960).

Salmonid production raceways (Figures 4.1 and 4.2) are usually constructed of poured reinforced concrete and generally have a length:width ratio of 10:1.

Buss and Miller (1971) recommended that raceway depths not exceed 2 ft (0.61 m), while Westers and Pratt (1977) recommended that raceways have a water exchange rate of 4 per hour and a linear velocity of 0.033 m/sec. An exchange rate of 4 per hour was deemed the best compromise between required pond size, velocity, and waste removal. A velocity greater than 0.033 m/sec did not allow solid waste to settle at the rear of the raceway and a lower velocity caused solids to settle within the pond, among the fish, making cleaning more difficult (Westers and Pratt 1977).

Raceways are generally built in series of multiple sections with a clean-out plug at the end of each section. The cleanouts are connected to a drainage line that leads to a settling pond. The clean-out drains are

Figure 4.1 Raceways for the commercial production of rainbow trout.

Figure 4.2 Raceways in South Carolina, used to produce trout to support recreational fishing.

located within a fish-free zone, called the quiescent zone (QZ), between a fish-retaining screen and the section discharge (Figure 4.3).

The number of sections that may be successfully placed in a series depends upon the reaeration capability between them (see Chapter 8) and the accumulation of dissolved metabolites (see Chapter 9).

Figure 4.3 Quiescent zone at the downstream end of a raceway section to collect solid waste produced by the fish above the screen.

The following example illustrates the process of a raceway design. Suppose a raceway is to be designed to receive 20 liters per second (L/sec) of water. The length:width ratio should be approximately 10:1, the desired depth is 0.2 m, the flow velocity must be 0.033 m/sec, and the exchange rate should be 4 per hour. The total volume of the raceway is

$$\frac{20\,L}{sec} \times \frac{3600\,sec}{hr} \times \frac{hr}{4\,exchanges} \times \frac{m^3}{1000\,L} = \frac{18\,m^3}{exchange}$$

Since the velocity and depth are set, the width must be calculated. Discharge (Q), velocity (V), and cross-sectional area (A) are related according to $Q = VA$ (Equation 3.1). The flow is set at 20 L/sec (0.02 m³/sec) and the depth is set at 0.2 m. Thus,

$$0.02\,m^3 = W \times 0.2 \times 0.033$$

and

$$W = \frac{0.02}{0.2 \times 0.03} = 30\,m$$

The total pond volume must be 18 m³, the depth is 0.2 m, and the width is 3.0 m. Thus, the length (L) is

$$L = \frac{18}{0.2 \times 0.03} = 30\,\text{m}$$

Notice that the requirement for a length:width ratio of 10:1 is met. An alternative raceway design method is to set the dimensions of the raceway at 30:3:1 (L:W:D). If a velocity of 0.033 m/sec for a 0.02 m^3/sec flow is desired, set D at x so that W = 3x. Then,

$$0.02 = 3x \times x \times 0.033,$$

$$x^2 = \frac{0.02}{3 \times 0.033}$$

$$x = 0.45\,\text{m} = D$$

$$W = 3x = 1.35\,\text{m}$$

A 10:1 length:width ratio is achieved by making the raceway 13.5 m in length. The total pond volume, then, is 18 m^3 and the exchange rate is 9 per hour. This raceway also has a satisfactory design and will have a greater fish density (weight of fish per unit of space) than the previous example. The preceding examples show that either depth, exchange rate, or dimensional ratios may be manipulated when designing a raceway with a particular velocity and dimensional requirements.

Buss et al. (1970) and Moody and McCleskey (1978) described the use of vertical raceways, or silos, in which water is injected at the bottom of a large vertical tube and allowed to exit at the top (Figure 4.4). Buss et al. (1970) first used 55-gal (208 L) steel drums with a water capacity of 43 gal (163 L). When the drums were supplied 3–6 gpm (11–23 L/min) of water, waste accumulated at the bottom. However, at 12 gpm (45 L/min), the units were self-cleaning. Self-cleaning drums were not advised if placed in a series because of the accumulation of suspended solid waste in downstream units. Buss et al. (1970) also reported successful trout production in fiberglass silos that were 16.5 ft (5.0 m) tall, 7.5 ft (2.3 m) in diameter, and supplied with 450 gpm (1703 L/min) of water. These units were self-cleaning so, presumably, a lower flow rate would be recommended if several were placed in a series. The water exchange rate in the silos was 5 per hour, which allowed fish densities to exceed those in typical trout raceways.

Moody and McCleskey (1978) described trout production in a serial silo system. The units were 11 ft (3.4 m) high, 7.25 ft (2.2 m) in diameter, and supplied with 300 gpm (1136 L/min) of water. Thus, the exchange rate was 5.3 per hour. There were five silos in the series with a 3 ft (0.9 m) drop between them and a clarifier between the fourth and fifth units.

Figure 4.4 Vertical raceway (silo) system described by Moody and McCleskey (1978) to produce rainbow trout in New Mexico. (Photo courtesy of Joe Kosalko.)

Fish growth in the silos was better than those grown in adjacent raceways, and fish densities were three times greater than in the raceways because of the increased rapid water exchange rate. Water quality was better through five successive uses in the silos than when reused similarly through raceways in series. Consequently, fish production was better in the silos than in the raceway series.

The major advantage of vertical raceways over horizontal units is that less physical space is required for the former (Buss et al. 1970). Additionally, fish feeding is probably less time-consuming in silos than in raceways (Buss et al. 1970; Moody and McCleskey 1978). A disadvantage of vertical units is the difficulty in fish removal for sampling, grading, or harvest. Vertical raceways are not in widespread use.

Circulating units

Circulating fish rearing units are usually circular in shape with a center drain. Water is admitted—usually under pressure—at a single location along the edge. Circular units are usually 2–3 ft (0.6–0.9 m) deep and 3–50 ft (0.9–15.2 m) in diameter (Figure 4.5).

Burrows and Chenoweth (1955) and Larmoyeux et al. (1973) studied the hydraulic characteristics of circular ponds. Both studies revealed a doughnut-shaped dead area in the center of the unit where water circulation was poor (Figure 4.6).

Figure 4.5 Circular fish rearing ponds used for the commercial production of rainbow trout in Montana.

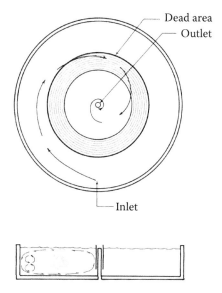

Figure 4.6 Hydraulic characteristics of a circular fish rearing unit. (After Burrows, R. E., and H. H. Chenoweth. 1955. *Progressive Fish-Culturist* 35: 122–131.)

Because of this dead area, mid-depth velocities were greatest at the edge, moderate near the outlet, and lowest in the center of the unit (Burrows and Chenoweth 1955). The size of the dead area and degree of stagnation can be reduced by increasing the depth of the pond or increasing the pressure at which water is injected (Burrows and Chenoweth 1955; Larmoyeux et al. 1973). Circular ponds are self-cleaning because the general decrease in velocity from the outside toward the center sweeps solid waste toward the drain. The doughnut-shaped stagnant zone is directly above an area with relatively high velocity, so that debris settling there is swept away (Burrows and Chenoweth 1955).

Losordo (1997) described a double-drain system for circular fish culture units that allows the separation of solid waste from the general flow of the unit (Figure 4.7).

In the double-drain system, settleable solids collect in a bowl at the bottom of the unit and are carried away in a separate stream from the liquid bulk, which exits from the surface of the unit. The center drain receives 15%–25% of the flow, with the remainder exiting through the sidewall discharge (Timmons and Ebeling 2013). This design may allow circular units to be used successfully for serial water reuse. Drawings of two types of double-drains are found in Losordo (1997) and Losordo et al. (1999).

Circulating square tanks with rounded corners (see Figure 11.9) are popular in Europe because they behave like round tanks, but take up less floor space. For this reason they may be selected for indoor recirculating aquaculture systems.

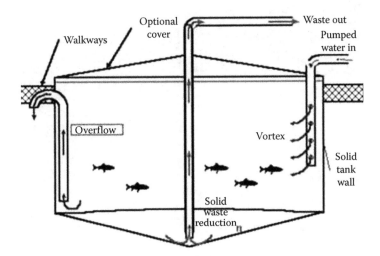

Figure 4.7 Double-drain circulating tank.

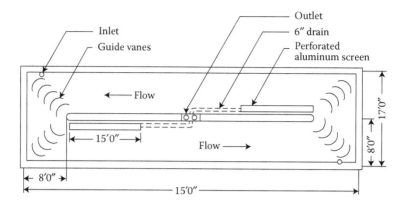

Figure 4.8 Rectangular, circulating pond. (From Burrows, R. E., and H. H. Chenoweth. 1970. *Progressive Fish-Culturist* 32: 67–80. With permission.)

Recommended tank diameter:depth ratios vary from 5:1 to 10:1 (Burrows and Chenoweth 1955; Larmoyeux et al. 1973). Timmons and Ebeling (2013) favor tanks with a diameter:depth ratio of less than 5:1. Circular tanks generally have water exchange rates not exceeding 2 per hour. Thus, fish densities in circular units are lower than in properly designed linear units.

Burrows and Chenoweth (1970) described a rectangular circulating pond with a center wall (Figures 4.8 and 4.9).

Figure 4.9 Burrows ponds used for the production of steelhead smolts at a National Fish Hatchery in Idaho.

Water is injected, under pressure, at two locations on opposite ends of the center wall. Drains are located at opposite ends of the center wall and are flush with the pond floor. The flow pattern around the pond is controlled by turning vanes positioned at each corner. Water flows parallel to the pond walls, and gradually reaches the center wall as velocity decreases. The pond is self-cleaning and contains no stagnant zones (Burrows and Chenoweth 1970). The so-called Burrows Pond has been used for the production of salmonid smolts. The water quality gradient of the linear sections of this hybrid design was thought to provide a healthier environment for fish rearing than circular ponds; the water velocity gradient provided by the circulating flow was thought to produce a stronger smolt that was better equipped for survival after release (Burrows and Chenoweth 1970) (Figure 4.8).

Circular ponds have flow velocities five to ten times greater than raceways (Burrows and Chenoweth 1955), which could be advantageous when rearing fish for increased stamina to improve survival after stocking. On the other hand, increased swimming speed increases oxygen demand, which lowers carrying capacity and increases food conversion rates, increasing the cost of production. Larmoyeux et al. (1973) reported that fish loads (weight of fish per unit of flow) in circular ponds could be higher than those in raceways receiving the same amount of water because dissolved oxygen is added to the unit by the high-pressure inflow. However, injecting water under pressure is not an energy-efficient means of aeration (see Chapter 8). Burrows and Chenoweth (1955) doubted that circular ponds could hold more fish than raceways receiving the same amount of water. Buss and Miller (1971) and Westers and Pratt (1977) reported that a water quality gradient, as provided by linear rearing units, was preferable to uniform water quality throughout the unit and recommended raceways over circular ponds for salmonid production.

A fundamental difference between raceway and circular pond designs is how solid waste is handled. Raceways collect a portion of the solids load in the QZs and water turbidity does not increase in successive downstream units if QZs are properly designed and fish densities are adequate (Soderberg 2007). For this reason, and due to the fact that raceways have minimal head loss, raceways are appropriate to serial reuse. Circular ponds are usually used for a single-water use because all of the solids exit the pond discharge and, thus, turbidity would accumulate in downstream units. The double-drain design partially solves this problem, but wastes up to 25% of system flow in the cleaning line. The head loss in a circular tank is equal to the depth, but can be decreased by using an external standpipe connected to the center drain. Grading and harvesting of fish is less labor-intensive for raceways than for circular ponds.

SAMPLE PROBLEMS

1. Design a raceway to receive 2400 L/min of water, have dimensional ratios of 30:3:1 (L:W:D) and have four water exchanges per hour.

2. What is the linear velocity of water in the unit you described above? Alter the raceway design so that the velocity is 0.033 m/sec and still provides four water exchanges per hour. Alter the raceway design so that the velocity is still 0.033 m/sec and the dimensional ratios are still 30:3:1. Which of the three raceway designs is best for fish culture? Why?

3. Raceways used for catfish production in Idaho are 7.3 m long, 2.4 m wide, and 0.9 m deep. There is one water exchange every 6 min. What is the flow rate (L/min) and velocity (m/sec)? How could the velocity be changes to 0.033 m/sec? What would be the new exchange rate?

4. What is the water exchange rate and vertical velocity of a 208 L steel drum with a water capacity of 163 L, supplied with 45 L/min of water? The drum diameter is 0.56 m.

5. Five water exchanges per hour are required for a vertical raceway that is 3 m in height and 2 m in diameter. How much water should be supplied to this unit?

6. Design a circular fish rearing unit to receive 30 L/sec of water.

7. Calculate the cost to construct the unit above if the walls and floor are 20 cm thick and reinforced concrete, poured in place, costs $200/m³.

8. Calculate the difference in construction cost between the unit you designed above and a raceway to receive the same flow.

9. If fish can be reared at 70 kg per L/sec of flow, compare fish densities, in kg/m³, between the raceway and the circular pond.

10. The rectangular circulating rearing pond (Burrows and Chenoweth 1970) is 23 m long, 5 m wide, and 0.8 m deep. What is the water exchange rate when the unit is supplied with 2300 L/min of water?

References

Burrows, R. E., and H. H. Chenoweth. 1955. *Evaluation of Three Types of Fish Rearing Ponds.* U.S. Fish and Wildlife Service Research Report 39.

Burrows, R. E., and H. H. Chenoweth. 1970. The rectangular circulating rearing pond. *Progressive Fish-Culturist* 32: 67–80.

Buss, K., D. R. Graff, and E. R. Miller. 1970. Trout culture in vertical units. *Progressive Fish-Culturist* 32: 187–191.

Buss, K., and E. R. Miller. 1971. Considerations for conventional trout hatchery design and construction in Pennsylvania. *Progressive Fish-Culturist* 33: 86–94.

Haskell, D. C., R. O. Davies, and J. Reckahn. 1960. Factors in hatchery pond design. *New York Fish and Game Journal* 7: 113–129.

Larmoyeux, J. D., R. G. Piper, and H. H. Chenoweth. 1973. Evaluation of circular tanks for salmonid production. *Progressive Fish-Culturist* 35: 122–131.

Losordo, T. M. 1997. Tilapia culture in intensive recirculating systems. Pages 185–211 *in* B. A. Costa-Pierce and J. E. Rakocy, editors. *Tilapia Aquaculture in the Americas*, Vol. 1. The World Aquaculture Society, Baton Rouge, Louisiana.

Losordo, T. M., M. B. Masser, and J. E. Rakocy. 1999. *Recirculating Aquaculture Tank Production Systems: A Review of Component Options*. Publication No. 453. Southern Regional Aquaculture Center, College Station, Texas.

Moody, T. M., and R. N. McCleskey. 1978. Vertical raceways for the production of rainbow trout. *New Mexico Department of Game and Fish Bulletin* No. 17.

Soderberg, R. W. 2007. Efficiency of trout raceway quiescent zones in controlling suspended solids. *North American Journal of Aquaculture* 69: 275–280.

Timmons, M. B., and J. M. Ebeling. 2013. *Recirculating Aquaculture*. Third edition. Publication No. 401-2013. Northeastern Regional Aquaculture Center, Ithaca, New York.

Westers, H., and K. M. Pratt. 1977. Rational design of hatcheries for intensive salmonid culture, based on metabolic characteristics. *Progressive Fish-Culturist* 39: 157–165.

chapter five

The solubility of oxygen in water

The gas laws

The gas laws, familiar to students of introductory chemistry, are of fundamental importance to fish culture and are therefore reviewed here. The combined gas law, which is a combination of Charles', Boyle's, and Gay-Lussac's laws, relates the volume of a gas to pressure and absolute temperature as follows:

$$\frac{V_1 P_1}{T_1} = \frac{V_2 P_2}{T_2}$$

According to this expression, the volume of a gas is proportional to its temperature and inversely proportional to the pressure exerted upon it.

Air is a mixture of gases (Table 5.1). Dalton's law states that the total pressure of a mixture of gases is equal to the sum of the individual partial pressures of its components. Standard pressure, or 1 atmosphere (atm), is 760 mm of mercury (mm Hg). Therefore, the partial pressure of oxygen (PO_2) in a total pressure of 1 atm of dry air is $760 \times 0.20946 = 159.2$ mm Hg. The atmosphere contains water vapor in addition to the gases listed in Table 5.1 which also exerts a partial pressure called vapor pressure (Pw). Saturation vapor pressures in relation to air temperature are listed in Table 5.2. The actual Pw is the saturation vapor pressure multiplied by the relative humidity, which is determined from the wet-bulb depression of a sling psychrometer. The PO_2 in moist air at a barometric pressure of P and a known Pw is $P - Pw \times 0.20946$. Hutchinson (1975) suggests that the saturation vapor pressure be subtracted from the vapor pressure when the relative humidity is more likely to be high than low. For routine aquaculture applications, correction of oxygen tensions for vapor pressure may be omitted except when accurate data are required for warm, humid locations.

Henry's law describes the solution of atmospheric gases in water. Simply stated, the equilibrium concentration of a gas in water is related to its partial pressure in the atmosphere above the water. The partial pressure of a gas dissolved in water, then, is the atmospheric pressure of that gas required to hold it in solution.

Table 5.1 Composition of dry air

Gas component	Percent of total volume
Nitrogen	78.084
Oxygen	20.946
Argon	0.934
Carbon dioxide	0.0407
Xenon, helium, neon, Methane, oxides of nitrogen	0.0026

Table 5.2 Saturation vapor pressures at different air temperatures

T (°C)	Pw (mm Hg)	T (°C)	Pw (mm Hg)
0	4.6	18	15.5
1	4.9	19	16.6
2	5.3	20	17.5
3	5.7	21	18.6
4	6.1	22	19.8
5	6.5	23	21.1
6	7.0	24	22.4
7	7.5	25	23.8
8	8.0	26	25.2
9	8.6	27	26.7
10	9.2	28	28.3
11	9.8	29	30.0
12	10.5	30	31.8
13	11.2	31	33.7
14	12.0	32	35.7
15	12.8	33	37.7
16	13.6	34	39.9
17	14.5	35	42.2

Calculation of the oxygen solubility in water

Chemists and physicists normally determine the solubility of dissolved gases using the Bunsen coefficient. Bunsen coefficients for oxygen are given in Table 5.3. To calculate the oxygen content (mg/L) of water in equilibrium with dry air at a pressure of 1 atm, the Bunsen coefficient is multiplied by the density of oxygen in mg/L, and then multiplied by 0.20946, the decimal fraction of oxygen in the air. For example, the solubility of oxygen in water at 10°C, in equilibrium with dry air at 1 atm is calculated as follows. The Bunsen coefficient is 0.03816 (Table 5.3).

Table 5.3 Bunsen coefficients and air solubilities of oxygen in moist air at 1.0 atm of pressure

Temperature	Bunsen coefficient	Solubility (mg/L)
0	0.04910	14.69
1	0.04777	14.24
2	0.04650	13.40
3	0.04529	13.40
4	0.04413	13.01
5	0.04302	12.64
6	0.04196	12.29
7	0.04095	11.95
8	0.03998	11.63
9	0.03905	11.31
10	0.03816	11.02
11	0.03730	10.73
12	0.03649	10.46
13	0.03570	10.20
14	0.03495	9.95
15	0.03423	9.71
16	0.03354	9.48
17	0.03288	9.26
18	0.03224	9.05
19	0.03163	8.85
20	0.03105	8.66
21	0.03048	8.47
22	0.02994	8.29
23	0.02942	8.12
24	0.02892	7.95
25	0.02844	7.80
26	0.02798	7.65
27	0.02754	7.36
28	0.02711	7.36
29	0.0267	7.22
30	0.0263	7.09
31	0.02592	7.00
32	0.02556	6.85
33	0.02521	6.73
34	0.02487	6.62
35	0.02455	6.51
36	0.02423	6.41

(*Continued*)

Table 5.3 (Continued) Bunsen coefficients and air solubilities of oxygen in moist
air at 1.0 atm of pressure

Temperature	Bunsen coefficient	Solubility (mg/L)
37	0.02393	6.31
38	0.02364	6.21
39	0.02336	6.12
40	0.02310	6.03

Source: Bunsen coefficients are from Colt, J. E. 1984. *Computation of Dissolved Gas Concentrations in Water as Functions of Temperature, Salinity, and Pressure.* American Fisheries Society Special Publication, 14. Bethesda, Maryland.

The volume of a mole of gas at 10°C is calculated using the combined gas law. The volume of a mole of gas at 0°C (273.15°K) is 22.4 L, according to Avogadro's law. Thus, the volume of a mole of gas at 10°C (283.15°K) is $22.4 \times 283.15/273.15 = 23.2$ L. The density of oxygen then, with a formula weight of 32, at 10°C, is

$$\frac{32,000\,\text{mg}}{\text{mole}} \times \frac{\text{mole}}{23.2\,\text{L}} = 1379.3\,\text{mg/L}$$

Thus, the solubility of oxygen with dry air at 10°C is $0.03816 \times 1379.3 \times 0.20946$, or 11.02 mg/L.

Tables of oxygen solubility have been presented by several authors (Winkler 1888; Roscoe and Lund 1889; Fox 1907, 1909; Truesdale et al. 1955). Mortimer (1956) reviewed these data and concluded that the results of Truesdale et al. (1955) are the most reliable, though they differ slightly from those calculated from the Bunsen coefficient (Table 5.3). The data of Truesdale et al. (1955) are recommended by Hutchinson (1975) and Boyd (1979) and are reproduced in (Table 5.4).

Most fish culturists use tables of oxygen solubility when this information is required; however, many workers find it more convenient and precise to calculate the oxygen solubility value for conditions at the time and place at which the value is needed.

Several authors provide empirical formulae for the determination of the equilibrium concentration of oxygen in water (Truesdale et al. 1955; Whipple et al. 1969; Liao 1971). The following regression expression (Truesdale et al. 1955) predicts the equilibrium concentration of oxygen in water in the temperature range of 0–36°C with a maximum deviation of 0.11 mg/L from their experimentally determined values (Table 5.4):

$$Ce = 14.161 - 0.3943\,T + 0.0077147\,T^2 - 0.0000646\,T^3 \qquad (5.1)$$

Table 5.4 Equilibrium concentrations of oxygen (Ce) with moist air at 1.0 atm pressure calculated from two empirical equations

T (°C)	Ce (mg/L)		Deviation (mg/L)
0	14.16	14.43	0.27
1	13.77	13.95	0.18
2	13.40	13.50	0.10
3	13.05	13.09	0.04
4	12.70	12.71	0.01
5	12.37	12.36	−0.01
6	12.06	12.03	−0.03
7	11.76	11.73	−0.03
8	11.47	11.44	−0.03
9	11.19	11.17	−0.02
10	10.92	10.92	0
11	10.67	10.68	0.01
12	10.43	10.45	0.02
13	10.20	10.24	0.04
14	9.98	10.04	0.06
15	9.76	9.85	0.09
16	9.56	9.66	0.10
17	9.37	9.49	0.12
18	9.18	9.32	0.14
19	9.01	9.16	0.15
20	8.84	9.01	0.17
21	8.68	8.86	0.18
22	8.53	8.72	0.19
23	8.38	8.59	0.21
24	8.25	8.46	0.21
25	8.11	8.34	0.23
26	7.99	8.22	0.23
27	7.86	8.10	0.24
28	7.75	7.99	0.24
29	7.64	7.88	0.24
30	7.53	7.78	0.25
31	7.43	7.68	0.25
32	7.33	7.58	0.25
33	7.23	7.49	0.26
34	7.13	7.40	0.27
35	7.04	7.31	0.27

Note: Column 2 was calculated using the equation of Truesdale et al. (1955) and Column 3 using Equation 5.2. (Modified after Liao, P. B. 1971. Water requirements of salmonids. *Progressive Fish-Culturist* 33: 210–215.) The last column shows the deviation from the values of Truesdale et al. (1955) when the simpler Equation 5.2 is used to calculate Ce.

where Ce = the equilibrium concentration of oxygen in water in mg/L with moist air at a pressure of 1 atm and T is the temperature in °C. This rather cumbersome equation can easily be programmed into a calculator.

Liao's (1971) formula for the calculation of Ce is Ce = $132/T^{0.625}$, where T is the temperature in °F. This formula is much easier to use than that of Truesdale et al. (1955) when a programmable calculator is not available, but the values obtained are 4%–8% higher than those determined by Truesdale et al. (1955). By changing the numerator, Liao's (1971) equation becomes

$$Ce = \frac{125.9}{T^{0.625}} \tag{5.2}$$

and predicts Truesdale's et al. (1955) experimental values with a maximum deviation of 0.27 mg/L (Table 5.4). This is sufficiently accurate for most fish culture applications.

Correction for pressure and salinity

The solubility of a gas in water varies with the atmospheric pressure of that gas in accordance with Henry's law. Therefore, the calculated value for Ce at standard pressure must be corrected for the atmospheric conditions at the location where the estimate is required. The most accurate way to correct for pressure is to measure the barometric pressure with a barometer when the oxygen solubility is to be calculated. The pressure correction factor, then, is

$$\frac{P}{760} \tag{5.3}$$

where P = the measured barometric pressure in mm Hg. If a barometer is not available, for general applications when average values are required, elevation can be used to determine the average atmospheric pressure. Liao (1971) provides the following pressure correction factor:

$$\frac{760}{760 + (E/32.8)} \tag{5.4}$$

where E = elevation in ft above sea level. In SI units, the pressure correction factor becomes

$$\frac{1}{1 + (E/7600)} \tag{5.5}$$

where E = elevation in m above sea level. The calculated oxygen solubility is multiplied by the pressure correction factor to obtain the pressure-corrected value.

Dissolved oxygen solubility is also influenced by the concentration of dissolved solids. Truesdale et al. (1955) provide a convenient equation to calculate the depression in Ce caused by salinity:

$$D_s = S(0.0841 - 0.00256\,T + 0.0000374\,T^2)$$

where D_s = reduction in Ce in mg/L due to salinity, S = salinity in parts per thousand (‰), and T = temperature in °C.

Measurement of dissolved oxygen concentration

Because of fish respiration, water used for aquaculture seldom contains the equilibrium concentration of dissolved oxygen (DO). The actual concentration of DO in water at any time is of fundamental concern to the fish culturist. Therefore, a convenient and accurate means of routinely determining the DO concentration is necessary for the operation of a fish hatchery.

The iodometric method developed by Winkler (1888), incorporating the azide modification to prevent nitrite interference, is widely used in water quality laboratories, but this procedure is too cumbersome for fish hatchery applications. A polarographic DO meter is much more useful and convenient for field applications and the results obtained are as accurate as those obtained from the Winkler method (Boyd 1979). Special care must be taken in purchasing an instrument from a reputable manufacturer, and in following the instructions carefully with regard to maintenance of the membrane and calibration of the instrument prior to each determination.

Expression of DO in tension units

The amount of DO in water is usually expressed in concentration units (mg/L), but for fish respiration considerations it is more usefully expressed as pressure in mm Hg. The oxygen tension of air (PO_2a) is the barometric pressure, minus the vapor pressure, multiplied by 0.20946. The tension of a given DO concentration (PO_2w) is the partial pressure of oxygen required to hold that amount of DO in solution. Thus, the PO_2w is the decimal fraction of oxygen saturation (measured DO divided by Ce), multiplied by the PO_2a above the water. If a barometer is not available, the average barometric pressure can be estimated from the elevation,

$$P = \frac{577,600}{760 + (E/32.8)} \tag{5.6}$$

where P = average barometric pressure in mm Hg at E ft above sea level, or,

$$P = \frac{760}{1 + (E/7600)} \tag{5.7}$$

where E = m above sea level.

SAMPLE PROBLEMS

1. The Bunsen coefficient for CO_2 at 20°C is 0.8705. Calculate the equilibrium concentration of CO_2 in water at this temperature and standard pressure. CO_2 has a formula weight of 44 g/mole. Note that the Bunsen coefficient for CO_2 is much larger than for O_2. The solubility is therefore greater. Why, then, is the equilibrium concentration of CO_2 less than for O_2?

2. Calculate the equilibrium concentrations of oxygen in the following waters:
   a. T = 20°C, E = 1500 ft above sea level
   b. T = 32°C, E = 200 m above sea level
   c. T = 10°C, P = 734 mm Hg
   d. Standard conditions: T = 20°C, P = 1 atm

3. Express the following oxygen concentrations as percent saturation and oxygen tension. Ignore the influence of vapor pressure.
   a. 5.0 mg/L at 10°C when P = 730 mm Hg
   b. 12 mg/L at 28°C when E = 250 m above sea level
   c. Saturated with oxygen under a pressure of 2.3 atm and a temperature of 8°C when released to the atmosphere at an elevation of 1000 m above sea level.
   d. Saturated with oxygen at 10°C, then heated to 20°C when P = 735 mm Hg

4. Calculate the altitude where half of the atmosphere by weight is below you and half is above you.

5. Calculate the equilibrium concentration of oxygen in a brackish water fish pond at sea level. T = 28°C and S = 17‰.

6. How much oxygen (kg) is in a cubic km of ocean water whose average temperature is 15°C? S = 35‰.

7. Calculate the DO concentration in 25°C water in a sealed plastic bag containing pure oxygen at a pressure of 1 atm.

8. Consider a DO concentration of 5.0 mg/L in 10°C water at an elevation of 500 m above sea level. If the air temperature is 25°C, how much does the PO_2w change when the relative humidity increases from 10% to 90%?

9. Calculate the percent error introduced by neglecting to correct the calculated PO_2w for Pw in the following examples. Assume the relative humidity is 100%.

    a. Air temperature 0°C, water temperature 8°C, water is saturated with DO, elevation is 350 m above sea level.

    b. Air temperature is 30°C, water temperature is 11°C, DO is 5.0 mg/L, elevation is 500 m above sea level.

    c. Air temperature is 30°C, water temperature is 30°C, water is saturated with DO at sea level.

    d. Air temperature is 35°C, water temperature is 28°C, DO is 3.5 mg/L, elevation is 50 m above sea level.

10. You have established that the lowest oxygen concentration resulting in acceptable growth of tilapia at 28°C is 3.5 mg/L. What DO concentration would be the same oxygen tension for salmon at 8°C?

References

Boyd, C. E. 1979. *Water Quality in Warmwater Fish Ponds*. Auburn University Agricultural Experiment Station, Auburn, Alabama.

Colt, J. E. 1984. *Computation of Dissolved Gas Concentrations in Water as Functions of Temperature, Salinity, and Pressure*. American Fisheries Society Special Publication, 14. Bethesda, Maryland.

Fox, C. J. J. 1907. On the coefficients of absorption of the atmospheric gasses in distilled water and in sea water. *Journal du Conseil/Conseil Permanent International pour l'Exploration de la Mer* 41.

Fox, C. J. J. 1909. On the coefficients of absorption of nitrogen and oxygen in distilled water and sea water, and of atmospheric carbonic acid in sea water. *Transactions of the Faraday Society* 5: 68–87.

Hutchinson, G. E. 1975. *A Treatise on Limnology: Chemistry of Lakes*, Vol 1, Part 2. John Wiley & Sons, New York, New York.

Liao, P. B. 1971. Water requirements of salmonids. *Progressive Fish-Culturist* 33: 210–215.

Mortimer, C. H. 1956. The oxygen content of air saturated fresh waters, and aids in calculating percentage saturation. *Mitteilung International Vereinigung fuer Theoretische unde Amgewundte Limnologie* 6: 1–20.

Roscoe, H. E. and J. Lund. 1889. On Schutzenberger's process for the estimation of dissolved oxygen in water. *Journal of the Chemical Society* 55: 552–576.

Truesdale, G. A., A. L. Downing, and G. F. Lowden. 1955. The solubility of oxygen in pure water and sea-water. *Journal of Applied Chemistry* 5: 53–62.

Whipple, W., Jr., J. V. Hunter, B. Davidson, F. Dittman, and S. Yu. 1969. *Instream Aeration of Polluted Rivers*. Water Resources Research Institute, Rutgers University, New Brunswick, New Jersey.

Winkler, L. W. 1888. Die loslichkeit des sauerstoffs im wasser. *Berichte der Deutschen Chemischen Gesellschuft* 22: 1764–1774.

chapter six

The oxygen requirements of fish

Fish respiration physiology

The fish gill apparatus is a marvelously functional structure for aquatic respiration. Each of four pairs of gill arches in teleosts contains posterior and anterior columns of gill filaments. Each filament bears transverse lamellae of a single cell layer of respiratory epithelium. As water is moved across the gill by the fish's bronchial pump, the filaments extend so that their tips touch those of adjacent filaments and the lamellae interdigitate with those on the filaments above and below. In this way, inspired water is forced through very small openings lined with respiratory tissue containing venous blood.

Fish respiratory efficiency is greatly aided by the fact that deoxygenated blood and oxygenated water flow along their contact area in opposite directions. In this way, water containing the most dissolved oxygen (DO) first comes in contact with blood partially loaded with oxygen. Farther down the area of blood-water interface, water with reduced oxygen tension contacts blood whose DO content is also low. Thus, the steepest possible oxygen tension gradient is maintained across the respiratory epithelium. Carbon dioxide (CO_2) transfer from blood to water is similarly favored by this mechanism.

Water for fish husbandry is normally not saturated with DO because of fish respiration. The response of fish to varying degrees of DO reduction is therefore of critical concern to the fish culturist. In respiration, fish blood picks up oxygen and releases CO_2 at the tissues. Therefore, the efficiency with which the blood combines with oxygen and CO_2 at different tensions determines the reaction of the fish to reduced DO levels in the water.

Oxygen dissociation curves (Figure 6.1) depict the extent to which blood is saturated with oxygen at different oxygen tensions. If the difference between the percentage of saturation of the hemoglobin between the tissues and the gill is great, much of the oxygen carried by the blood will be released at the tissues as illustrated by the sigmoid curve (Figure 6.1). However, little oxygen will be released at the tissues when the difference between the percentages of saturation at oxygen tensions existing at the two locations is small. This situation is depicted by the hyperbolic oxygen dissociation curve (Figure 6.1). The shapes of oxygen dissociation curves are influenced by the species of fish (Fry 1957), temperature (Fry 1957;

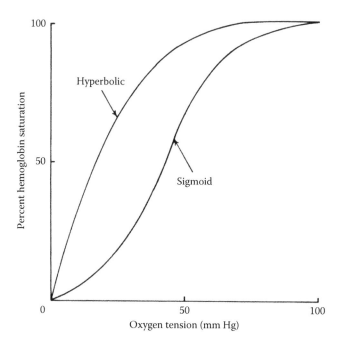

Figure 6.1 Model hyperbolic and sigmoid fish oxygen dissociation curves.

Cameron 1971), and the tension of PCO_2 (Black et al. 1966; Cameron 1971). The tendency is for the oxygen dissociation curve to become hyperbolic (less efficient oxygen delivery to the tissues) with decreasing temperature and increasing PCO_2.

Oxygen consumption rates of cultured fish

The ability to predict the amount of oxygen that fish will extract from the water under different conditions will allow us to calculate the weight of fish a water stream of known discharge and oxygen content should support. Various authors (Brett 1962; Beamish 1964a,b,c; Beamish and Dickie 1967; Brett and Zala 1975; Caulton 1976) have studied the oxygen consumption rates of individual fish in respiration chambers. From this research we know that the amount of DO that fish must extract from water flowing past them depends upon their size, sex, reproductive cycle, activity level, temperature, and the oxygen content of the water. Oxygen consumption of confined fish is simpler than in nature because the size of the fish in a lot is relatively uniform, the sex ratio is usually one, cultured fish are generally not allowed to reach sexual maturity, the activity level is constant, and water temperatures and DO concentrations are constant or fluctuate predictably.

Elliot (1969) studied the oxygen consumption rate of chinook salmon between 1.85 and 17.50 g in weight. He measured the DO concentrations above and below groups of fish under hatchery conditions and developed an empirical method for estimating their rates of oxygen consumption. For fish from 1.85 to 5.90 g,

$$Yn = [0.02420T - 0.7718] - [(0.001242T - 0.04544)(n - 1.85)] \quad (6.1)$$

and for fish from 5.90–17.5 g,

$$Yn = [0.01917T - 0.5877] - [(0.0003676T - 0.011601)(n - 5.90)] \quad (6.2)$$

where Yn = mg/L of DO per lb of fish per gpm of flow (n = average individual fish weight in g and T = temperature in °F).

Liao (1971) collected oxygen consumption data for trout and Pacific salmon in hatcheries and developed an equation that predicts oxygen uptake when water temperature and average fish size are known:

$$O_2 = KT^nW^m \quad (6.3)$$

where O_2 = oxygen consumption rate in weight units of DO per 100 weight units of fish per day, K = rate constant, T = temperature in °F, W = average individual fish weight in lbs and m and n are slopes. The following constants were provided:

Species	T (°F)	K	m	n
Salmon	≤50	7.20×10^{-7}	−0.194	3.200
Salmon	>50	4.90×10^{-5}	−0.194	2.120
Trout	≤50	1.90×10^{-6}	−0.138	3.130
Trout	>50	3.05×10^{-4}	−0.138	1.855

Liao's (1971) formula gives nearly identical values as Elliot's (1969) for the oxygen consumption rate of Pacific salmon (Figure 6.2) and the two may be used interchangeably for fish smaller than 17.5 g.

Muller-Fuega et al. (1978) conducted an experiment with rainbow trout under hatchery conditions similar to Liao's (1971) investigation. Much like Liao (1971), they found a discontinuity in oxygen consumption versus temperature at 10°C (50°F). Their expression predicting oxygen consumption rate for rainbow trout is

$$OD = ap^{\beta}10^{gt} \quad (6.4)$$

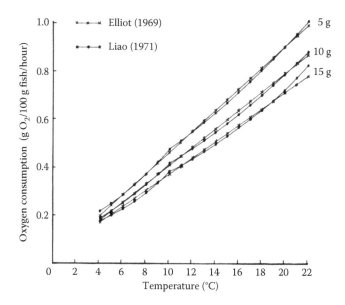

Figure 6.2 Oxygen consumption rates of 5, 10, and 15 g Pacific salmon at different water temperatures using the equations of Elliot (1969) and Liao (1971).

where OD = mg oxygen demand in mg DO/kg of fish per hour, p = weight of individual fish in g, T = temperature in °C and α, β and γ are constants. The constant values are as follows:

	Temperature (°C)	
Constant	4–10	12–22
α	75	249
β	−0.196	−0.142
γ	0.055	0.024

The results obtained from equations by Liao (1971) and Muller-Fuega et al. (1978) are identical at temperatures greater than 12°C, but different at temperatures less than 12°C (Figures 6.3 through 6.5). This difference is considerable and we shall refer to Willoughby's (1968) extensive examination of hatchery carrying capacities for further information.

Willoughby (1968) related oxygen consumption to the amount of food fed to trout in hatcheries with the following expression:

$$(Oa - Ob) \times 0.0545 \times gpm = lbs \text{ of food per day}$$

where Oa = DO in mg/L at the pond influent, Ob = DO in mg/L at the pond effluent, and gpm = pond discharge in gpm for a diet containing

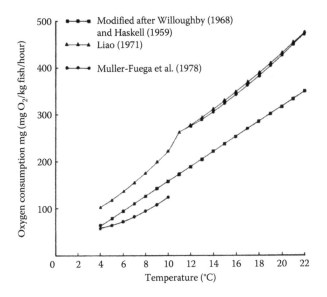

Figure 6.3 Oxygen consumption rates of 50 g trout calculated from Equation 6.5, based on the expressions of Haskell (1959) and Willoughby (1968) and from the equations of Liao (1971) and Muller-Fuega et al. (1978).

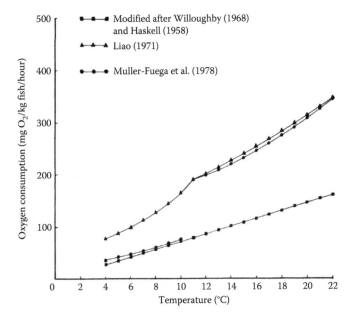

Figure 6.4 Oxygen consumption rates of 250 g trout calculated from Equation 6.5, based on the expressions of Haskell (1959) and Willoughby (1968) and from the equations of Liao (1971) and Muller-Fuega et al. (1978).

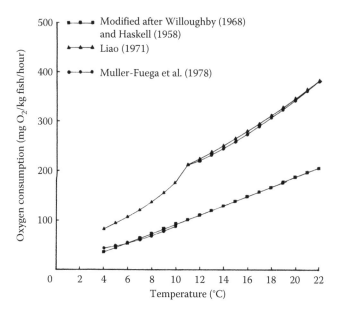

Figure 6.5 Oxygen consumption rates of 500 g trout calculated from Equation 6.5, based on the expressions of Haskell (1959) and Willoughby (1968) and from the equations of Liao (1971) and Muller-Fuega et al. (1978).

1200 C/lb. When Haskell's (1959) feeding equation (Equation 2.2) is applied to Willoughby's (1968) formula for a given temperature and size of fish, the oxygen consumption rate may be calculated. This is reasonable since both Liao (1971) and Muller-Fuega et al. (1978) collected data on fish being fed at rates based on Haskell's (1959) equation. The two expressions thus combined is

$$Oc = \left(\frac{3 \times C \times \Delta L}{L} \right) \times (9155.23) \qquad (6.5)$$

where Oc = oxygen consumption in mg DO/kg of fish per hour, C = food conversion, ΔL = daily increase in length, and L = fish length. Data from this equation, using a food conversion of 1.7 and a growth rate of 5.9 Centigrade temperature units per cm, fall midway between the results from Liao (1971) and Muller-Fuega (1978) in the lower temperature range and about 25% lower for the higher temperature range (Figure 6.3). Of the three methods available for estimating the oxygen consumption rates of trout in culture, Equation 6.5, based on Haskell's (1959) and Willoughby's (1968) work, may be preferable because it allows for variable growth, food conversion, and diet types to be taken into account.

Andrews and Matsuda (1975) presented oxygen consumption data for channel catfish of different sizes and at several temperatures. Boyd et al. (1978) applied multiple regression analysis to these data to obtain the following expression:

$$\log O_2 = -0.999 - 0.000957\ W + 0.0000006\ W^2$$
$$+ 0.0327\ T - 0.0000087\ T^2 + 0.0000003\ WT \qquad (6.6)$$

where O_2 = oxygen consumption rate in mg DO/g of fish per hour, W = average individual fish weight in g and T = temperature in °C. This formula may be used to predict the oxygen consumption rate for any size of catfish at any temperature.

Ross and Ross (1983) presented regression equations for the resting oxygen consumption rates of tilapia, *Oreochromis niloticus*. OC = oxygen consumption rate in mg, oxygen/kg, fish/hr, and wt = fish body weight in g.

Temperature (°C)	Regression equation	
20	log OC = 3.00–0.777 (log wt)	(6.7_{20})
25	log OC = 2.80–0.350 (log wt)	(6.7_{25})
30	log OC = 3.34–0.586 (log wt)	(6.7_{30})
35	log OC = 3.03–0.255 (log wt)	(6.7_{35})

Effects of hypoxia on growth

The relationship between the oxygen consumption rate and the environmental oxygen tension has been investigated by Fry and Hart (1948), Graham (1949), Shepherd (1955), Holeton and Randall (1967), and Itazawa (1970). The classical explanation given by Fry (1957) is that oxygen consumption is constant and at a maximum at DO tensions above the incipient limiting level. The incipient lethal level is the tension under which mortality results from prolonged exposure (Figure 6.6).

Most workers agree that between the incipient limiting and lethal levels of DO, the oxygen consumption rate is dependent upon environmental DO. This zone of respiratory dependence should also be a zone of reduced growth since less than the maximum amount of oxygen is delivered to the tissues. Holeton and Randall (1967) found that the oxygen consumption rate of rainbow trout did not change over a range of DO tensions from 40–160 mm Hg. The fish increased their ventilation volume to keep a constant level of oxygen supplied to the tissues.

Whether the mechanism of reduced DO limiting fish growth is decreased oxygen consumption or the osmoregulatory cost of hyperventilation is not particularly germane to practical fish culture. The degree

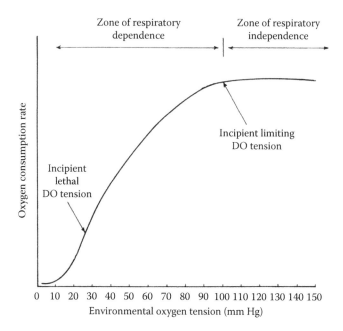

Figure 6.6 Oxygen consumption by fish as related to environmental oxygen tension. (After Fry, F. E. J. 1957. *The Physiology of Fishes*, Vol. 1. Academic Press, New York, New York.)

to which growth is reduced by various amounts of DO reduction is, however, of primary concern because a DO minimum for culture water must be defined designating at which point the water must be reaerated or discharged.

Brungs (1971), in a carefully controlled long-term study, found that the growth of fathead minnows (*Pimephales promelas*) was reduced at all DO concentrations below saturation. Andrews et al. (1973) reported that channel catfish ate less and grew more slowly at 60% DO saturation (approximately 95 mm Hg) than at 100% saturation. Carlson et al. (1980) found that at 25°C, channel catfish growth was reduced at a constant DO exposure of 3.5 mg/L (69 mm Hg), but there was no growth reduction at 5.1 mg/L (100 mm Hg). Larmoyeaux and Piper (1973) reared rainbow trout in a system where the water flowed through a series of troughs, each containing fish. Growth was reduced in the trough where the DO had fallen to 4.2 mg/L (61 mm Hg) by fish respiration upstream, but not in the trough above it where the DO was 4.9 mg/L (71 mm Hg). Ammonia exposure (see Chapter 9) may have contributed to this reduction in growth. Downey and Klontz (1981) found that rainbow trout growth was not different at 105.3mm Hg and 138.5 mm Hg, but growth was reduced by 10% at 81.5 mm Hg. Weight gain at 70 mm Hg was 66.6% of that at 128 mm Hg.

Assignment of DO minima for hatcheries

Willoughby (1968), Piper (1970), Smith and Piper (1975), Leitritz and Lewis (1976), and Westers and Pratt (1977) suggested that aquaculture facilities for trout be designed so that fish are exposed to a minimum DO concentration of 5.0 mg/L. Buss and Miller (1971) called for aeration at trout hatcheries when the DO level was predicted to fall below 5.0–7.0 mg/L. Burrows and Combs (1968) recommended a DO minimum of 6.0 mg/L for Pacific salmon. Because fish respiration occurs along oxygen tension gradients, criteria for DO minima for fish culture are more usefully and universally expressed in tension units. Furthermore, use of tension units allows changes in oxygen solubility under different conditions to be taken into account.

Reported incipient limiting DO tensions for freshwater fish range from 60 mm Hg for brown bullheads (*Ictaluris nebulosus*) (Grigg 1969), to 100 mm Hg for brook trout (Graham 1949) and rainbow trout (Itazawa 1970). Davis (1975) conducted a statistical analysis of available data on the incipient limiting data and recommended minimum PO_2 levels of 86 mm Hg for mixed fish populations not including salmonids and 90 mm Hg for salmonid populations (Table 6.1). Downey and Klontz (1981) recommended minimum oxygen tensions of 90 mm Hg for trout in hatcheries.

Table 6.1 Reported incipient limiting DO tensions and recommended DO minima

Species	Incipient limiting DO (mm Hg)	Reference
Brown bullhead	60	Grigg (1969)
Common carp *Cyprinus carpio*	80	Itazawa (1970)
Rainbow trout	78	Irving et al. (1941)
Rainbow trout	100	Itazawa (1970)
Rainbow trout	80	Cameron (1971)
Brook trout	78	Irving et al. (1941)
Brook trout	100	Graham (1949)
Tilapia (*O. niloticus*)	60	Ross and Ross (1983)
Tilapia (*O. niloticus*)	26–31	Becker and Fishelson (1986)
Recommended	DO minima (mm Hg)	
Fish communities without salmonids	73	Davis (1975)
Fish communities with salmonids	86	Davis (1975)
Salmonid fish communities	90	Davis (1975)

Ross and Ross (1983) determined the incipient limiting DO tension (referred to as pC, critical point) to be 60 mm Hg (2.92 mg/L at 28°C) for tilapia (*O. niloticus*). Becker and Fishelson (1986) reported values of 26–31 mm Hg (28.5 mm Hg = 1.39 mg/L at 28°C) for the incipient limiting level of DO for *O. niloticus*. Field observations on DO concentrations below which tilapia growth is reduced range from 2.3 mg/L (Coche 1977) to 3.0 mg/L (Melard and Philppart 1980). Thus, the value of 60 mm Hg appears to be a more reasonable value for assignment of DO minima for flowing water culture of tilapia.

These recommended DO minima may be unnecessarily conservative for practical fish culture. Since the incipient limiting DO is the point where oxygen-dependent growth begins, DO minima based on this value should provide for maximum growth. Perhaps some growth reduction, in exchange for the increased carrying capacity that a lower DO minimum would provide, is reasonable in aquaculture. As previously cited, several trout hatchery design specialists have recommended a DO minimum of 5.0 mg/L. With 10°C water at an elevation of 1500 m above sea level, this concentration corresponds to a PO_2 of 73 mm Hg.

Respiratory efficiency is related to the steepness of the oxygen tension gradient between water and blood. Thus, warmwater fish should be able to tolerate lower concentrations of DO than fish in cold water. This is because the tension of a given concentration of oxygen becomes greater as the equilibrium concentration decreases. If a DO minimum of 80 mm Hg for trout is accepted, assigning DO minima of the same PO_2 for other species, whose oxygen requirements have not been studied, is reasonable. For example, catfish raised at sea level in 28°C water would be exposed to 3.9 mg/L DO when the PO_2 was 80 mm Hg.

SAMPLE PROBLEMS

1. Calculate the oxygen consumption rates for fish in the following situations using Liao's formula (Equation 6.3)
   a. Salmon at 8°C, fish weight is 102 g each
   b. Trout at 8°C, each fish is 22 cm long
   c. Salmon at 11°C, fish weight is 42 g each
   d. Trout at 15°C, each fish is 25 cm long
2. Repeat Problem 1 using Muller-Fuega's formula (Equation 6.4).
3. Repeat Problem 1 using Equation 6.5. Convert to common units to compare your results from Problems 1, 2, and 3.
4. 100 50 g catfish are placed in a concrete tank that is 1.3 m long, 0.6 m wide, and 0.3 m deep. The water temperature is 28°C and is saturated with oxygen. How long will it take for the fish to reduce the PO_2 to 50 mm Hg? Use Equation 6.6. Repeat these calculations using Equation 6.5. Compare results.

5. A 5-Ha catfish pond, 1 m deep, contains 4500 300 g fish per Ha. The water temperature is 27°C. An aeration system delivers 0.6 kg of oxygen per kWh. How large a unit (how many kW) will be required to replace the oxygen in the pond removed by the fish? Use Equations 6.5 and 6.6.

6. 500 g catfish are reared in a raceway receiving water geothermally heated to 29°C. What is their oxygen consumption rate? Use Equations 6.5 and 6.6.

7. You have established that 5.0 mg/L is a reasonable DO minimum for trout in 10°C water at an elevation of 300 m above sea level. On this basis, what minimum DO level would you assign for catfish grown at 27°C and 50 m above sea level?

8. Compare the amount of DO, in mg/L, available for respiration at the following two locations if influent DO is 100% of saturation and effluent DO is 75 mm Hg.

   Location 1: T = 11°C, E = 200 m above sea level
   Location 2: T = 30°C, E = 200 m above sea level

9. 500,000 8-cm trout are placed in a raceway that receives 2300 L/min of 12°C water. The elevation of the site is 500 m above sea level and the oxygen content of the water entering the raceway is 100% of saturation. What will the DO and PO_2 be at the tail end of the raceway? Use Equation 6.5.

10. Describe the conflict in fish hatchery management between fish quality (healthy fish that are able to survive hatchery conditions and resist disease) and fish quantity (weight of fish per available unit of water flow). Based on the literature cited in this chapter, what minimum oxygen level would you assign for fish hatcheries?

References

Andrews, J. W., T. Murai, and G. Gibbons. 1973. The influence of dissolved oxygen on the growth of channel catfish. *Transactions of the American Fisheries Society* 102: 835–838.

Andrews, J. W. and Y. Matsuda. 1975. The influence of various culture conditions on the oxygen consumption of channel catfish. *Transactions of the American Fisheries Society* 104: 322–327.

Beamish, F. W. H. 1964a. Seasonal changes in the standard rate of oxygen consumption of fishes. *Canadian Journal of Zoology* 42: 189–194.

Beamish, F. W. H. 1964b. Respiration of fishes with special emphasis on standard oxygen consumption. II. Influence of weight and temperature on respiration of several species. *Canadian Journal of Zoology* 42: 177–188.

Beamish, F. W. H. 1964c. Respiration of fishes with special emphasis on standard oxygen consumption. III. Influence of oxygen. *Canadian Journal of Zoology* 42: 355–366.

Beamish, F. W. H. and L. M. Dickie. 1967. Metabolism and biological production in fish. Pages 215–242 *in* S. D. Gerking, editor. *The Biological Basis for Freshwater Fish Production*. Blackwell, Oxford, England.

Becker, K. and L. Fishelson. 1986. Standard and routine metabolic rate, critical oxygen tension and spontaneous scope for activity of tilapias. Pages 623–628 *in* J.L. Maclean, L.B. Dizon and L.V. Hosillos editors. *Proceedings of the First Asian Fisheries Forum*. Asian Fisheries Society, Manila, Philippines.

Black, E. C., D. Kirkpatrick, and H. H. Tucker. 1966. Oxygen dissociation curves of the blood of brook trout, *Salvelinus fontinalis*, acclimated to summer and winter temperatures. *Journal of the Fisheries Research Board of Canada* 23: 1–13.

Boyd, C. E., R. P. Romaire, and E. Johnston. 1978. Predicting early morning dissolved oxygen concentrations in channel catfish ponds. *Transactions of the American Fisheries Society* 107: 484–492.

Brett, J. R. 1962. Some considerations in the study of respiratory metabolism in fish, particularly salmon. *Journal of the Fisheries Research Board of Canada* 19: 1025–1038.

Brett, J. R. and C. A. Zala. 1975. Daily pattern of nitrogen excretion and oxygen consumption of sockeye salmon, *Oncorhynchus nerka*, under controlled conditions. *Journal of the Fisheries Research Board of Canada* 32: 2479–2486.

Brungs, W. A. 1971. Chronic effects of low dissolved oxygen concentrations on the fathead minnow, *Pimephales promelas*. *Journal of the Fisheries Research Board of Canada* 28: 1119–1123.

Burrows, R. E. and B. D. Combs. 1968. Controlled environments for salmon propagation. *Progressive Fish-Culturist* 30: 123–136.

Buss, K. and E. R. Miller. 1971. Considerations for conventional trout hatchery design and construction in Pennsylvania. *Progressive Fish-Culturist* 33: 86–94.

Cameron, J. N. 1971. Oxygen dissociation characteristics of the blood of rainbow trout, *Salmo gairdneri*. *Comparative Biochemistry and Physiology* 38A: 699–704.

Carlson, A. R., J. Blocher, and L. J. Herman. 1980. Growth and survival of channel catfish and yellow perch exposed to lowered constant and diurnally fluctuating dissolved oxygen concentrations. *Progressive Fish-Culturist* 42: 73–78.

Caulton, M. S. 1976. The effect of temperature on the routine metabolism in *Tilapia rendalli* Boulenger. *Journal of Fish Biology* 11: 549–553.

Coche, A. G. 1977. Premiers resultats de l'elevage en cages de *Tilapia nilotica* (L.) dans le Lac Kossou, Cote d'Ivoire. *Aquaculture* 10: 109–140.

Davis, J. C. 1975. Minimal dissolved oxygen requirements of aquatic life with special emphasis on Canadian species: A review. *Journal of the Fisheries Research Board of Canada* 32: 2295–2332.

Downey, P. C. and G. W. Klontz. 1981. *Aquaculture Techniques: Oxygen (PO2) Requirements for Trout Quality*. Idaho Water and Energy Resources Research Institute. University of Idaho, Moscow, Idaho.

Elliot, J. W. 1969. The oxygen requirements of chinook salmon. *Progressive Fish-Culturist* 31: 67–73.

Fry, F. E. J. 1957. The aquatic respiration of fish. Pages 1–63 *in* M. E. Brown, editor. *The Physiology of Fishes*, Vol. 1. Academic Press, New York, New York.

Fry, F. E. J. and J. S. Hart. 1948. Cruising speed of goldfish in relation to water temperature. *Journal of the Fisheries Research Board of Canada* 7: 169–175.

Graham, J. M. 1949. Some effects of temperature and oxygen pressure on the metabolism and activity of speckled trout, *Salvelinus fontinalis*. *Canadian Journal of Research* 27: 270–288.

Grigg, G. C. 1969. The failure of oxygen transport in a fish at low levels of ambient oxygen. *Comparative Biochemistry and Physiology* 29: 1253–1257.

Haskell, D. C. 1959. Trout growth in hatcheries. *New York Fish and Game Journal* 6: 204–237.

Holeton, G. F. and D. J. Randall. 1967. The effect of hypoxia upon the partial pressure of gasses in the blood and water afferent and efferent to the gills of rainbow trout. *Journal of Experimental Biology* 46: 317–327.

Irving, L., E. C. Black, and V. Stafford. 1941. The influence of temperature upon the combination of oxygen with blood of trout. *Biological Bulletin* 80: 1–17.

Itazawa, Y., 1970. Characteristics of respiration of fish considered from the arterio-venous difference of oxygen content. *Bulletin of the Japanese Society of Scientific Fisheries* 36: 571–577.

Larmoyeaux, J. D. and Piper, R. G. 1973. The effects of water reuse on rainbow trout in hatcheries. *Progressive Fish-Culturist* 35: 2–8.

Leitritz, E. and R. C. Lewis. 1976. *Trout and Salmon Culture—Hatchery Methods*, State of California, Department of Fish and Game, Fish Bulletin 164.

Liao, P. B. 1971. Water requirements of salmonids. *Progressive Fish-Culturist* 33: 210–215.

Melard, C. H. and J. C. Philppart. 1980. Pisciculture intensive de *Sarotherodon niloticus* (L.) dans les effluents thermiques d'une centrale nucléaire en Belgique. *EIFAC Symposium on New Developments in Utilization of Heated Effluents and of Recirculation Systems for Intensive Aquaculture*. Stavanger, Norway. EIFAC/80/DOC.E/11.

Muller-Fuega, A., J. Petit, and J. J. Sabaut. 1978. The influence of temperature and wet weight on the oxygen demand of rainbow trout, *Salmo gairdneri* R., in fresh water. *Aquaculture* 14: 355–363.

Piper, R. G. 1970. Know the proper carrying capacities of your farm. *American Fishes and U.S. Trout News* 15: 4–6.

Ross, B. and L.G. Ross. 1983. The oxygen requirements of *Oreochromis niloticus* under adverse conditions. Pages 134–143 *in* L. Fishelson and Z. Yaron, editors. *Proceedings of the First International Symposium on Tilapia in Aquaculture*. Tel Aviv University, Tel Aviv, Israel.

Shepherd, M. P. 1955. Resistance and tolerance of young speckled trout, *Salvelinus fontinalis*, to oxygen lack, with special reference to low oxygen acclimation. *Journal of the Fisheries Research Board of Canada* 12: 387–433.

Smith, C. E. and R. G. Piper. 1975. Lesions associated with chronic exposure to ammonia. Pages 497–514 *in* W. E. Ribelin and G. Migaki, editors. *The Pathology of Fishes*. University of Wisconsin Press, Madison, Wisconsin.

Westers, H. and K. M. Pratt. 1977. Rational design of hatcheries for intensive salmonid culture, based on metabolic characteristics. *Progressive Fish-Culturist* 39: 157–165.

Willoughby, H. 1968. A method for calculating carrying capacities of hatchery troughs and ponds. *Progressive Fish-Culturist* 30: 173–174.

chapter seven

Rearing density and carrying capacity

Independence of loading rate and rearing density

Tunison (1945) presented recommended densities for trout in troughs and ponds and demonstrated that acceptable densities increase with increasing fish size. Haskell (1955) related permissible fish density to feeding rate, recognizing that it was a function of oxygen consumption and metabolic production, and that these parameters were proportional to the amount of food metabolized by the fish.

The authors recommended practical fish densities in weight of fish per volume unit of the fish rearing container, but it is obvious that permissible fish densities should be related to water flow because the volume of flow determines the amount of oxygen available for respiration and the degree of metabolite dilution. Buss et al. (1970) and Piper (1975) described the interdependence of water requirements and spatial requirements of cultured fish. It is now conventional terminology to refer to the weight of fish per water flow unit as the loading rate and the weight of fish per volume unit of rearing space as density. The maximum permissible loading rate is that which results in effluent DO at the predetermined minimum allowable tension. This is often referred to as carrying capacity. The maximum permissible density may be determined by the hydraulic characteristics of the rearing unit or by the physical, physiological, or behavioral spatial requirements of the fish.

The important fish husbandry parameters of carrying capacity and maximum density are independent of each other and presented separately here.

Calculation of hatchery carrying capacity

Willoughby's method

Based on Haskell's (1955) principle that oxygen consumption and metabolite production were proportional to the amount of food fed, Willoughby

(1968) presented a formula that related oxygen consumption to the food ration:

$$(Oa - Ob) \times 0.0545 \times gpm = \text{pounds of food per day}$$

where Oa = DO of incoming water in mg/L, Ob = effluent DO, and gpm = volume of flow in gal/min. The feed rate in percent body weight per day is obtained from Haskell's (1959) formula. Rearranging Willoughby's formula and inserting the feed rate expression,

$$\text{Loading rate} = \frac{(Oa - Ob)(0.0545)}{F} \tag{7.1}$$

where loading rate = lb fish/gpm and F = feeding rate in weight of food per weight of fish per day. When the value for Ob is the minimum allowable DO, the loading rate is at the maximum safe level; Equation 7.1 solves for the carrying capacity.

Piper's method

Haskell's (1959) feed rate formula demonstrates that the feed rate is proportional to fish length. Because the permissible weight of fish in a rearing unit is directly related to the feed rate (Haskell 1959), it follows that the carrying capacity should be related to fish length. Recognizing this, Piper (1975) introduced the concept of Flow Index: F = W/L, where F = flow index, W = lb of fish/gpm, and L = fish length in inches. Solving for carrying capacity,

$$\text{Carrying capacity} = F \times L \tag{7.2}$$

Piper (1975) stated that the desirable flow index for his station in Bozeman, Montana was 1.5, but recognized that this was a site-specific criterion related to the DO content of the water. Cannady (Piper et al. 1982) developed a table (Table 7.1) that shows recommended flow indices for various conditions of available DO as a function of temperature and elevation, if influent water is saturated with DO. The values in Cannady's table are based on a desirable flow index of 1.5 at 5000 ft (1524 m) above sea level and 50°F (10°C) water (the conditions at the Bozeman station). For example, the recommended flow index for 48°F (8.9°C) water at an elevation of 2000 ft (610 m) above sea level is 1.85 (Table 7.1) if influent water is saturated with oxygen.

If the water supplied to a rearing unit is not saturated with DO, such as in the case of raceways in series, Piper et al. (1982) recommended

Table 7.1 Cannady's table to relate flow index to temperature and elevation

Water temperature (°F)	Elevation, ft above sea level									
	0	1000	2000	3000	4000	5000	6000	7000	8000	9000
40	2.70	2.61	2.52	2.43	2.34	2.25	2.16	2.09	2.01	1.94
41	2.61	2.52	2.44	2.35	2.26	2.18	2.09	2.02	1.94	1.87
42	2.52	2.44	2.35	2.27	2.18	2.10	2.02	1.95	1.88	1.81
43	2.43	2.35	2.27	2.19	2.11	2.03	1.94	1.88	1.81	1.74
44	2.34	2.26	2.18	2.11	2.03	1.95	1.87	1.81	1.74	1.68
45	2.25	2.18	2.10	2.03	1.95	1.88	1.80	1.74	1.68	1.61
46	2.16	2.09	2.02	1.94	1.87	1.80	1.73	1.67	1.61	1.55
47	2.07	2.00	1.93	1.86	1.79	1.73	1.66	1.60	1.54	1.48
48	1.98	1.91	1.85	1.78	1.72	1.65	1.58	1.53	1.47	1.42
49	1.89	1.83	1.76	1.70	1.64	1.58	1.51	1.46	1.41	1.36
50	1.80	1.74	1.68	1.62	1.56	1.50	1.44	1.39	1.34	1.29
51	1.73	1.67	1.62	1.56	1.50	1.44	1.38	1.34	1.29	1.24
52	1.67	1.61	1.56	1.50	1.44	1.39	1.33	1.29	1.24	1.19
53	1.61	1.55	1.50	1.45	1.39	1.34	1.29	1.24	1.20	1.15
54	1.55	1.50	1.45	1.40	1.34	1.29	1.24	1.20	1.16	1.11
55	1.50	1.45	1.40	1.35	1.30	1.25	1.20	1.16	1.12	1.07
56	1.45	1.40	1.35	1.31	1.26	1.21	1.16	1.12	1.08	1.04
57	1.41	1.36	1.31	1.27	1.22	1.17	1.13	1.09	1.05	1.01
58	1.36	1.32	1.27	1.23	1.18	1.14	1.09	1.05	1.02	0.98
59	1.32	1.28	1.24	1.19	1.15	1.10	1.06	1.02	0.99	0.95
60	1.29	1.24	1.20	1.16	1.11	1.07	1.03	0.99	0.96	0.92
61	1.25	1.21	1.17	1.13	1.08	1.04	1.00	0.97	0.93	0.87
62	1.22	1.18	1.14	1.09	1.05	1.01	0.97	0.94	0.91	0.87
63	1.18	1.14	1.11	1.07	1.03	0.99	0.95	0.92	0.88	0.85
64	1.15	1.12	1.08	1.04	1.00	0.96	0.92	0.89	0.86	0.83

Source: From Piper, R. G. et al. 1982. *Fish Hatchery Management.* U.S. Department of the Interior, Fish and Wildlife Service, Washington, DC.

decreasing the flow index in proportion to the decrease in available DO. For example, a site with a DO equilibrium concentration of 10 mg/L and a desired DO minimum of 4.7 mg/L has 10 – 4.7 = 5.3 mg/L of DO available for fish respiration. If the influent DO is 8 mg/L, 8 – 4.7 = 3.3 mg/L is available and the selected flow index from Table 7.1 should be multiplied by 3.3/4.7, or 0.70.

Calculation of carrying capacity from oxygen consumption

Knowledge of the oxygen consumption rate of fish allows for direct calculation of their water requirement. Empirical formulae for oxygen consumption rates are available for Chinook salmon (Elliot 1969; Liao 1971), trout (Willoughby 1968; Liao 1971; Muller-Fuega et al. 1978), catfish (Boyd et al. 1978), and tilapia (Ross and Ross 1983) (see Chapter 6). Derivation of the required unit conversions follow. Let CC = carrying capacity in kg/L/min, Ci = influent DO, and Cm = minimum allowable DO so that Ci – Cm = the DO available for respiration in mg/L.

For Chinook salmon (Elliot 1969),

$$CC\left(\frac{lb}{gpm}\right) = \frac{(Ci-Cm)mg}{L} \times \frac{lb\,fish}{Yn\,mg/L \times gpm}$$

and

$$CC = \frac{Ci-Cm}{Yn} \times \frac{kg}{2.2lb} \times \frac{gpm}{3.78L/min}$$

and

$$CC = \frac{0.12(Ci-Cm)}{Yn} \tag{7.3}$$

where Yn = the oxygen consumption rate in (mg/L)/(lb fish/gpm) calculated from Equation 6.1 or 6.2. For Pacific salmon and trout (Liao 1971),

$$CC = \frac{(Ci-Cm)mg}{L} \times \frac{100\,mg\,fish/day}{Oc\,mg} \times \frac{kg}{10^6\,mg} \frac{1440\,min}{day}$$

and

$$CC = \frac{0.144(Ci-Cm)}{Oc} \tag{7.4}$$

where Oc = the oxygen consumption rate in weight units DO/100 weight units of fish/day) calculated from Equation 6.3.

For trout (Muller-Fuega et al. 1978),

$$CC = \frac{(Ci - Cm)mg}{L} \times \frac{kg\,fish/hr}{OD\,mg} \times \frac{60\,min}{hr}$$

and

$$CC = \frac{60(Ci - Cm)}{OD} \qquad (7.5)$$

where OD = the oxygen consumption rate in mg/kg fish/hr calculated from Equation 6.4.
    For catfish (Boyd et al. 1978),

$$CC = \frac{(Ci - Cm)mg}{L} \times \frac{g\,fish/hr}{O_2\,mg} \times \frac{60\,min}{hr}$$

and

$$CC = \frac{0.06(Ci - Cm)}{OC} \qquad (7.6)$$

where O_2 = oxygen consumption rate in mg DO/g fish/hr from Equation 6.5.
    For tilapia (*Oreochromis niloticus*) (Ross and Ross 1983),

$$CC = \frac{(Ci - Cm)mg}{L} \times \frac{g\,fish/hr}{OC} \times \frac{kg}{1,000g} \times \frac{60\,min}{hr}$$

and

$$CC = \frac{0.06(Ci - Cm)}{OC} \qquad (7.7)$$

where OC = oxygen consumption rate in mg oxygen/kg fish/hour from Equation 6.6.

Comparison of methods for calculation of carrying capacity

Direct calculation of carrying capacity is necessary for species other than trout and Pacific salmon because indirect methods have not been described

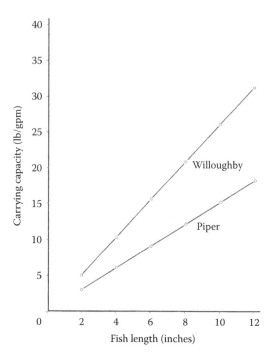

Figure 7.1 Trout carrying capacity as a function of fish length calculated from Willoughby (1968) and Piper (1975). Assumptions used are flow index = 1.5, elevation = 1500 m above sea level, influent water saturated with DO, DO minimum = 75 mm Hg.

for them. Three authors (Willoughby 1968; Liao 1971; Muller-Fuega et al. 1978) provide methods for determining the oxygen consumption rates of trout and Elliot (1969) and Liao (1971) describe similar procedures for Pacific salmon (see Chapter 6). All methods give slightly different results, but Willoughby's (1968) method may be preferable because it allows for variable growth, food conversion, and diet composition to be considered. Piper's (1975) flow index method is much simpler to use than calculating carrying capacity from the oxygen consumption rate, but gives substantially lower values (Figures 7.1 and 7.2).

Fish rearing density

From the preceding sections, it is evident that the water requirements of fish are fairly well known. Their spatial requirements, however, are rather poorly defined. A fish rearing unit is designed to receive a particular volume of water flow. Thus, the hydraulic requirements of the rearing

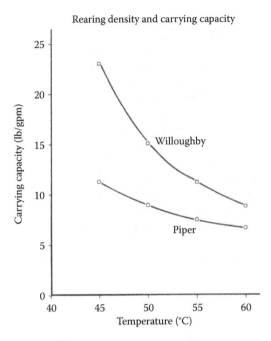

Figure 7.2 Carrying capacity of 15-cm trout at different temperatures calculated from Willoughby (1968) and Piper (1975). Assumptions: Piper—flow index is for 1500 ft above sea level from Table 7.1; Willoughby—growth rate = 7.1 MTU/cm, C = 1.5, elevation = 1500 ft above sea level, influent water saturated with DO, DO minimum = 75 mm Hg.

container set the maximum fish density because carrying capacity must not be exceeded. Westers (1970) relates carrying capacity to fish density

$$lb/ft^3 = \frac{lb/gpm \times R}{8} \tag{7.8}$$

where R = the number of water exchanges per hour. This useful equation relates fish density and loading rate per the hydraulic characteristics of the rearing unit, but does not address the spatial requirements of fish.

Spatial requirements of cultured fish

The behavioral aspects of fish spatial requirements can be addressed by the example in which individuals of an aggressive species such as bluegill (*Lepomis macrochirus*) are successively added to an aquarium. Adverse

effects are not noted until the fifth fish is added because there are four dis-
tinct territories (corners) to defend in the aquarium. As more individuals
are added, this territoriality breaks down and behavior no longer limits
the density at which fish may be held. Even low hatchery rearing densities are great enough to preclude behavioral limitations, with the possible
exception of Atlantic salmon.

Some references infer that crowding increases the incidence of
infectious disease and fin erosion. Soderberg and Krise (1986) reported
greater mortality due to bacterial disease in lake trout reared at 15.4 lb/ft^3
(270 kg/m^3) than in those reared at lower densities, up to 9.2 lb/ft^3 (148 kg/
m^3). Their experimental rearing containers were constructed from rough
plastic netting that may have caused abrasions allowing epizootics to
occur in the higher density treatments. Buss et al. (1970) reared rainbow
trout at 34 lb/ft^3 (545 kg/m^3) in smooth-sided containers without disease
incidence (Figure 7.3).

A correlation between fin erosion and rearing density in Atlantic
salmon has been reported by Westers and Copeland (1973) and
Maheshkumar (1985), but such an effect was not noted in other studies

Figure 7.3 Trout reared at extreme density in a hatching jar. Spatial considerations did not affect fish performance if sufficient water flow for respiration was
maintained. (Photo courtesy of Keen Buss.)

with lake trout (Soderberg and Krise 1987) and Atlantic salmon (Schneider and Nicholson 1980; Soderberg and Meade 1987; Soderberg et al. 1993a).

Density index

Piper (1975) related recommended fish density to fish length in the term Density Index. The Density Index is simply the allowable density in lb/ft^3 divided by the fish length in inches. A Density Index of 0.5 was recommended for trout production in hatcheries (Piper 1975). Maximum rearing density is calculated as

$$\text{Rearing density} = L \times 0.5 \qquad (7.9)$$

where rearing density is in lb/ft^3 and L is fish length in inches. Thus, 8-inch (20.3 cm) trout could be reared safely at 4 lb/ft^3 (64 kg/m^3) as long as their water requirements for respiration and metabolite dilution were met.

Density Index has been widely adopted for salmonid hatchery design, but trout have been reared at densities considerably greater than those common at most hatcheries. Clary (1979) found that trout were not adversely affected when reared at a Density Index of 0.8. Soderberg and Krise (1986) noted no effects on 7.7-inch (19.6-cm) lake trout reared at a Density Index of 1.0 (7.7 lb/ft^3; 124 kg/m^3). In a similar study (Soderberg et al. 1987), 4.7-inch (11.9-cm) lake trout were successfully reared at density indices as high as 2.0 (9.4 lb/ft^3; 151 kg/m^3). These results indicate that allowable fish densities may not be related to fish length and that trout may be safely reared at densities of around 10 lb/ft^3 (160 kg/m^3) regardless of their size. Because small fish have greater water requirements than larger fish, their high-density culture requires greater flows than equal densities of larger fish.

Atlantic salmon appear to have different spatial requirements than other salmonid species. Keenleyside and Yamamoto (1962) reported that Atlantic salmon must maintain discrete feeding territories for optimum growth and survival. Fenderson and Carpenter (1971) observed that Atlantic salmon hatchery production is related to their weight per unit area of the rearing container rather than its volume, because they orient to the bottom surface rather than occupying the entire water column. Early recommendations on rearing densities range from 0.75 lb/ft^2 (3.7 kg/m^2) of rearing unit floor surface (Piggins 1971) to 1.5 lb/ft^2 (7.3 kg/m^2)(Peterson et al. 1972). However, Westers and Copeland (1973) found that growth did not decrease until densities approached 3.0 lb/ft^2 (14.7 kg/m^2). Soderberg and Meade (1987) reared Atlantic salmon at densities up to 4.3 lb/ft^2 (21.0 kg/m^2) in 8°C water and Soderberg et al. (1993b) achieved densities up to 2.86 lb/ft^2 (14.0 kg/m^2) at 17.5°C without observing adverse effects on growth or survival.

Schmittou (1969) successfully reared channel catfish in cages suspended in ponds at densities as high as 11.9 lb/ft³ (191 kg/m³). Fish growth, feed conversion, and survival were not affected by density in the range of 7.2–11.9 lb/ft³ (115–191 kg/m³). Thus, a maximum density for catfish in cages was not demonstrated (Schmittou 1969). Ray (1979) determined that catfish could be reared in raceways at a density of at least 10 lb/ft³ (160 kg/m³). Hickling (1971) reported that common carp (*Cyprinus carpio*) are reared in Japan at densities as high as 17.4 lb/ft³ (278 kg/m³).

The factors that determine practical densities for intensive fish culture are not known, but the evidence presented here indicates that crowding alone does not limit salmonid production at rearing densities two to three times greater than are usually experienced in hatchery conditions. The consequence of this conclusion to salmonid hatchery design is that most facilities provide more rearing space than is necessary for their production requirements. Rearing space at existing hatcheries could be reduced to decrease production costs and rearing units at proposed hatcheries can be made smaller than those typically used for salmonid production.

SAMPLE PROBLEMS

1. Three trout raceways in series are supplied with water that is saturated with oxygen. Thus, the influent DO to the first unit is the equilibrium concentration. Use Piper's formula (Equation 7.2) and the equation modified from Willoughby's method (Equation 7.1) to calculate carrying capacity of each of the three units assuming that aeration between each unit restores DO to 80% of saturation.

2. How many 400 g catfish could you raise in a 30 m³/sec discharge from a power plant that averages 28°C? The minimum allowable DO tension is 75 mm Hg. What pond volume would be required if the maximum rearing density is 160 kg/m³?

3. A trout hatchery in Pennsylvania has a flow of 10,000 L/min at 9°C. How many 15-cm trout could be held there at one time? How many linear feet of raceway would be required if they are 2 m wide and 0.6 m deep?

4. What is the carrying capacity of 25-cm trout at a North Carolina farm that receives 1200 L/min of 15°C water at an elevation of 1000 m above sea level?

5. Brackish groundwater with a temperature of 28°C and a salinity of 10% is supplied to a tilapia farm in Kenya. How much water is required to grow 100,000 kg of 450 g fish?

6. Prepare a graph that compares the results of Equations 7.1 and 7.2 for calculating trout carrying capacity to direct calculation from the oxygen consumption rate using Liao's equation (Equation 7.4) and

Muller-Fuega's equation (Equation 7.5). How are they similar? How are they different?

7. How many 20-cm Atlantic salmon smolts could be reared in a 1700 L/min water supply at sea level at 11°C? Assume one water use.

8. Calculate the difference in carrying capacity of trout fed a 2600 C/kg diet and a 3300 C/kg diet.

9. How many 15-cm catfish should be stocked in a cylindrical cage 1.0 m in diameter and 1.2 m deep if their density is to be 200 kg/m³ when they reach a size of 35 cm?

References

Boyd, C. E., R. P. Romaire, and E. Johnston. 1978. Predicting early morning dissolved oxygen concentrations in channel catfish ponds. *Transactions of the American Fisheries Society* 107: 484–492.

Buss, K., D. R. Graff, and E. R. Miller. 1970. Trout culture in vertical units. *Progressive Fish-Culturist* 32: 187–191.

Clary, J. R. 1979. High density trout culture. *Salmonid* 2: 8–9.

Elliot, J. W. 1969. The oxygen requirements of Chinook salmon. *Progressive Fish-Culturist* 31: 67–73.

Fenderson, O. E. and M. R. Carpenter. 1971. Effects of crowding on the behavior of juvenile hatchery and wild Atlantic salmon (*Salmo salar*). *Animal Behavior* 19: 439–447.

Haskell, D. C. 1955. Weight of fish per cubic foot of water in hatchery troughs and ponds. *Progressive Fish-Culturist* 17: 117–118.

Haskell, D. C. 1971. Trout growth in hatcheries. *New York Fish and Game Journal* 6: 204–237.

Hickling, C. F. 1971. *Fish Culture*. Faber and Faber, London, England.

Keenleyside, M. H. A. and F. T. Yamamoto. 1962. Territorial behavior of juvenile Atlantic salmon (*Salmo salar* L.). *Behavior* 19: 139–169.

Liao, P. B. 1971. Water requirements of salmonids. *Progressive Fish-Culturist* 33: 210–215.

Maheshkumar, S. 1985. The epizootiology of finrot in hatchery-reared Atlantic salmon (*Salmo salar*). Master's thesis. University of Maine, Orono, Maine.

Muller-Fuega, A., J. Petit, and J. J. Sabaut. 1978. The influence of temperature and wet weight on the oxygen demand of rainbow trout, *Salmo gairdneri* R., in fresh water. *Aquaculture* 14: 355–363.

Peterson, H. H., O. T. Carlson, and S. Johansson. 1972. *The Rearing of Atlantic Salmon*. Astro-Ewos AB, Sodertalje, Sweden.

Piggins, D. J. 1971. Smolt rearing, tagging and recapture techniques in a natural river system. *International Atlantic Salmon Foundation Special Publication Series* 2: 63–82.

Piper, R. G. 1975. *A Review of Carrying Capacity Calculations for Fish Hatchery Rearing Units*. Information Leaflet No 1, U.S. Fish and Wildlife Service Fish Culture Development Center, Bozeman Fish Technology Center, Bozeman, Montana.

Piper, R. G., I. B. McElwain, L. E. Orme, J. P. McCraren, L. G. Fowler, and J. R. Leonard. 1982. *Fish Hatchery Management*. U.S. Department of the Interior, Fish and Wildlife Service, Washington, DC.

Ray, L. 1979. Channel catfish production in geothermal water. Pages 192–195 *in* L. J. Allen and E. C. Kinney, editors. *Proceedings of the Bio-Engineering Symposium for Fish Culture.* Publication 1, Fish Culture Section, American Fisheries Society, Bethesda, Maryland.

Ross, B. and L. G. Ross. 1983. The oxygen requirements of *Oreochromis niloticus* under adverse conditions. Pages 134–143 *in* L. Fishelson and Z. Yaron, editors. *Proceedings of the First International Symposium on Tilapia in Aquaculture.* Tel Aviv University: Tel Aviv, Israel.

Schmittou, H. R. 1969. Developments of the culture of channel catfish, *Ictalurus punctatus* R., in cages suspended in ponds. 23rd Annual Conference of the Southeastern Association of Game and Fish Commissoners, Mobile, Alabama.

Schneider, R. and B. L. Nicholson. 1980. Bacteria associated with finrot disease in hatchery-reared Atlantic salmon (*Salmo salar*). *Canadian Journal of Fisheries and Aquatic Sciences* 37: 1505–1513.

Soderberg, R. W., D. S. Baxter, and W. F. Krise. 1987. Growth and survival of fingerling lake trout reared at four different densities. *Progressive Fish-Culturist* 49: 284–285.

Soderberg, R. W. and W. F. Krise. 1986. Effects of density on growth and survival of lake trout, *Salvelinus namaycush. Progressive Fish-Culturist* 48: 30–32.

Soderberg, R. W. and W. W. Krise. 1987. Fin condition of lake trout, *Salvelinus namaycush* Walbaum, reared at different densities. *Journal of Fish Diseases* 10: 233–235.

Soderberg, R. W. and J. W. Meade. 1987. Effects of rearing density on growth, survival and fin condition of Atlantic salmon. *Progressive Fish-Culturist* 49: 280–283.

Soderberg, R. W., J. W. Meade, and L. A. Redell. 1993a. Fin condition of Atlantic salmon reared at high densities in heated water. *Journal of Aquatic Animal Health* 5: 77–79.

Soderberg, R. W., J. W. Meade and L. A. Redell. 1993b. Growth, survival and food conversion of Atlantic salmon reared at four different densities with common water quality. *Progressive Fish-Culturist* 55: 29–31.

Tunison, A. V. 1945. *Trout Feeds and Feeding.* Cortland Experimental Hatchery, Cortland, New York. Mimeo.

Westers, H. 1970. Carrying capacity of salmonid hatcheries. *Progressive Fish-Culturist* 32: 43–46.

Westers, H. and J. Copeland. 1973. *Atlantic Salmon Rearing in Michigan, Michigan Department of Natural Resources, Fisheries Division*, Technical Report 73-27, Lansing, Michigan.

Willoughby, H. 1968. A method for calculating carrying capacities of hatchery troughs and ponds. *Progressive Fish-Culturist* 30: 173–174.

chapter eight

Reaeration of flowing water

When fish respiration has reduced the dissolved oxygen (DO) tension to the minimum allowable level, water must be reconditioned by aeration to be of further use for fish production. The extent to which water may be reused with aeration as the only treatment measure depends upon the accumulation of toxic metabolic by-products. A detailed description of this fish husbandry parameter follows in Chapter 9.

Aeration theory

Transfer of oxygen into water is a three-stage process in which gaseous oxygen is transferred to the surface film, diffuses through the surface film, and then finally moves into the liquid bulk by convection. Since oxygen enters water by diffusion, the rate of oxygen transfer depends upon the area of air–water interface and the oxygen deficit of the water. Diffusion of atmospheric oxygen in aquaculture systems where water is quiescent or moving in laminar flow and oxygen deficits are quite small is too slow to be an important source of DO for fish respiration unless the area of air–water interface is increased by turbulence or agitation. Aeration of water streams used for aquaculture can be accomplished by gravity where the energy released when the water loses altitude is used to increase the air–water surface, or by mechanical devices that spray water into the air or inject air or pure oxygen into the water.

Gravity aeration devices

The most logical means of improving oxygen regimes of cultured fish is by gravity fall of water between production units that is provided by the topography at the facility (Figures 8.1 and 8.2). The extent to which water is reaerated by gravity is of fundamental concern for practical fish culture in flowing water.

Haskell et al. (1960) compared aeration by water passage over a simple weir (Figures 8.3a and 8.4) with aeration obtained by water flow over a splashboard (Figures 8.3b and 8.5) that broke the waterfall part way down and then flowed over various screen and slat arrangements at the dam. Chesness and Stephens (1971) evaluated several devices for increasing

Figure 8.1 Raceways on a level site provide little elevational difference between units.

Figure 8.2 Two views of an aquaculture site where topography allows for considerable gravity aeration between production units.

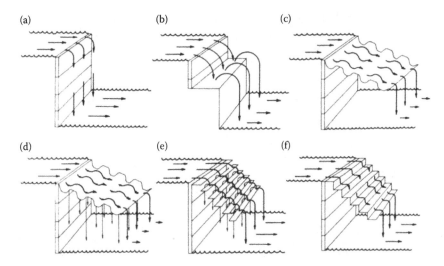

Figure 8.3 Diagrams of gravity aerators. (a) Simple weir (Haskell et al. 1960; Chesness and Stephens 1971); (b) splashboard (Haskell et al. 1960; Chesness and Stephens 1971); (c) inclined corrugated sheet (Chesness and Stephens 1971); (d) inclined corrugated sheet with holes (Chesness and Stephens 1971); (e) lattice aerator (Chesness and Stephens 1971); and (f) cascade aerator (Tebbutt 1972). (Adapted from Soderberg, R. W. 1982. *Progressive Fish-Culturist*, 44: 89–93. With permission.)

Figure 8.4 Water flowing from one raceway to another over a simple weir.

Figure 8.5 Gravity aeration accomplished with splashboards placed between ponds.

oxygen transfer over a gravity fall including a splashboard, an inclined sheet of corrugated roofing material (Figure 8.3c), a similar corrugated sheet pierced with holes (Figures 8.3d and 8.6), and an open stairstep device referred to as a lattice (Figure 8.3e). Tebbutt (1972) studied aeration down stairstep arrangements called cascades (Figure 8.3f).

Figure 8.6 Gravity aerator built with corrugated roofing material pierced with holes. (Photo courtesy of R. O. Smitherman.)

The following equation (Downing and Truesdale 1955) can be used to evaluate and compare aeration devices:

$$E = 100 \times \frac{\text{actual increase in DO}}{\text{possible increase in DO}}$$

or

$$E = 100 \times \frac{Cb - Ca}{Ce - Ca} \tag{8.1}$$

where E = efficiency of the aeration device, Cb = DO in mg/L below the device, Ca = DO above the device, and Ce = equilibrium concentration of DO at the aeration site. Selected data on measured efficiencies of some gravity aerators over various distances of waterfall are presented in Table 8.1.

Table 8.1 Selected data on measured efficiencies of some gravity aerators over various distances of waterfall

Device	Waterfall (cm)	Efficiency (%)
Simple weir	22.9[a]	6.2
	30.5[b]	9.3
Inclined corrugated sheet[b]	30.5	25.3
	61.0	43.0
Inclined corrugated sheet with holes[b]	30.5	30.1
	61.0	50.1
Splashboard	22.9[a]	14.1
	30.5[b]	24.1
	61.0[b]	38.1
Lattice[b]	30.5	34.0
	61.0	56.2
Cascade[c]	25.0	23.0
	50.0	33.4
	75.0	41.2
	100.0	52.4

Source: From Soderberg, R. W. 1982. *Progressive Fish-Culturist* 44: 89–93. With permission.

[a] Haskell et al. (1960).
[b] Chesness and Stephens (1971).
[c] Tebbutt (1972).

To solve for the expected DO below an aeration device of known efficiency, the efficiency equation can be rearranged as

$$Cb = \frac{E(Ce - Ca)}{100} + Ca \qquad (8.2)$$

Suppose there is a 30-cm drop between two ponds with a simple weir separating them. If the water temperature is 10°C and the elevation is 183 m above sea level, the solubility of oxygen is 10.67 mg/L (Chapter 4). If the fish loading in the upstream pond is such that the DO concentration is depressed to 5.0 mg/L, Ca will be 5.0. We know that a 30-cm fall over a simple weir is 9.3% efficient (Table 8.1). The DO below the weir, then, is

$$Cb = \frac{9.3(10.67 - 5.0)}{100} + 5.0 = 5.53 \text{ mg/L}$$

Notice that this procedure compensates for the oxygen deficit of water being aerated and for a given amount of increase in the air–water interface provided by a particular aeration device; the actual oxygen transfer is related to the deficit Ce – Ca. In the previous example, 5.53 – 5.0, or 0.53 mg/L of DO, was added to the water. If this water is sent over the aeration device a second time,

$$Cb = \frac{9.3(10.67 - 5.53)}{100} + 5.53 = 6.01 \text{ mg/L}$$

and the oxygen transferred is 6.01 – 5.53, or 0.48 mg/L, because the oxygen deficit of the water being aerated is lower.

Although the benefit of gravity aeration can be considerable, most aquaculture sites require mechanical aeration or liquid oxygen to realize the production potential of their water supplies.

Mechanical aeration devices

Mechanical units that agitate the water surface are commonly used in flowing water aquaculture systems because of their convenience and ease of installation (Figure 8.7). Aerators are evaluated and compared based on their ability to transfer oxygen to water. Tests are conducted under standard conditions of 760 mm Hg pressure, 20°C temperature, and zero DO in the water being aerated. The amount of oxygen added to the water in a given amount of time under a certain power level is measured. The rating in kg of oxygen per shaft kW per hr (kg/kWh) or lb of oxygen per HP per hr (lb/HPh) is given by the aerator manufacturer as a measure of its efficiency and can be used to compare units. Actual oxygen transfer depends

Figure 8.7 Mechanical aerator commonly used in flowing water aquaculture.

upon the oxygen deficit because as saturation is approached an increasing amount of power is required per unit of DO transferred. Reaeration above 95% saturation can seldom be justified on a cost basis (Mayo 1979). Westers and Pratt (1977) list 90% saturation as a reasonable design criterion for reaerated water. Since aquaculture systems operate at the relatively high DO minima of 3 to 7 mg/L, actual transfer rates will be lower than those determined under standard conditions.

Surface aerators are generally rated to transfer 1.9 to 2.3 kg/kWh under standard conditions (Eckenfelder 1970). Whipple et al. (1969) found that mechanical aerators in polluted rivers generally provided oxygen transfer rates of 0.61 kg/kWh or less, but their test water was higher in oxygen demand than is usual for aquaculture effluents. Soderberg et al. (1983) reported an average transfer rate of 0.83 kg/kWh in static water trout ponds where fish were heavily fed and aeration began when DO tensions reached 75 mm Hg. Aeration of flowing water should be more efficient than in static pools because processed water is continually replaced from upstream rather than being recirculated around the unit. To estimate aeration requirements, a conservative value such as 0.6 kg/kWh may be used or the oxygen transfer may be estimated from

$$RT = RS \frac{(\beta Ce_t - Ca)(1.025^{T-20})(\alpha)}{Ce_{20}}$$

where RT = actual oxygen transfer, RS = oxygen transfer under standard condition, Ce_t = oxygen solubility at the aeration site, Ca = DO concentration of the water being aerated, T = temperature in °C, Ce_{20} = oxygen solubility under standard conditions, β = DO solubility in aerated water/DO solubility in clean water, and α = DO transfer in aerated water/DO transfer in clean water. Since aquaculture waters are generally relatively unpolluted, values of 0.85 and 1.0 may be used for α and β, respectively. The solubility of DO at 20°C and 760 mm Hg pressure is 8.84 mg/L (see Chapter 4), thus, the oxygen transfer equation reduces to

$$RT = RS\frac{(Ce_t - Ca)(1.025^{T-20})(0.85)}{8.84} \tag{8.3}$$

The following example illustrates the use of this formula. Suppose an aerator is rated by the manufacturer to transfer oxygen at 2.0 kg/kWh under standard conditions. The actual oxygen transfer at a site where the water temperature is 10°C, barometric pressure is 725 mm Hg, and the DO of the water to be aerated is 5.0 mg/L, may be estimated as follows:

RS = 2.0 kg/kWh
Ce_t = 10.42 mg/L (see Chapter 4)
Ca = 5.0 mg/L
T = 10°C

$$RT = 2.0\frac{(10.42 - 5.0)(1.025^{10-20})(0.85)}{8.84} = 0.81 \text{ kg/kWh}$$

When the actual oxygen transfer rate (RT) has been estimated, the aeration capability of a particular unit and set of conditions can readily be obtained. For the above example, suppose that a 1.0 kW unit is placed in a water flow of 4000 L/min. The DO concentration below the aerator (Cb) is

$$Cb = Ca + \frac{0.81 \text{ kg}}{kW \text{ h}} \times \frac{10^6 \text{ mg}}{kg} \times 1.0 \text{ kW} \times \frac{\text{min}}{4000 \text{ L}} \times \frac{\text{hr}}{60 \text{ min}}$$

and

$$Cb = 5.0 + 3.38 = 8.38 \text{ mg/L}$$

Similarly, an aerator may be sized for a particular job. Suppose that, for the above example, a unit that will return the DO to 90% saturation is desired. The amount of oxygen required is 0.9 (10.42)–5.0 = 4.38 mg/L,

$$\frac{4.38 \text{ mg}}{L} \times \frac{kg}{10^6 \text{ mg}} \times \frac{4000 \text{ L}}{\text{min}} \times \frac{60 \text{ min}}{\text{hr}} = 1.05 \text{ kg/hr}$$

The size of the unit required in kW of shaft power is

$$\frac{1.05 \text{ kg}}{\text{hr}} \times \frac{kW \text{ h}}{0.81 \text{ kg}} = 1.3 \text{ kW}$$

Burrows and Combs (1968) describe an aeration system appropriate for aerating an entire hatchery water supply (Figures 8.8 and 8.9).

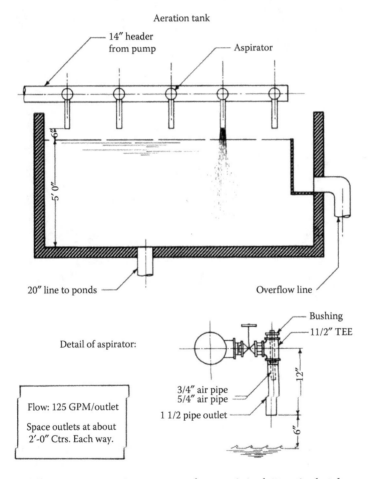

Figure 8.8 Burrows-type aspirator system for aeration of an entire hatchery water supply. (From Burrows, R. E. and Combs, B. D. 1968. *Progressive Fish-Culturist* 30: 133. With permission.)

Figure 8.9 Two views of a Burrows type aspirator at a large salmonid hatchery in Idaho. The adjacent tanks are biofilters to remove ammonia so that the water can be reused (see Chapter 11).

Diffuser aeration devices

Diffused air systems introduce air or oxygen into the water. The effectiveness of oxygen transfer is obviously a function of bubble size (as it affects the area of air–water interface), the oxygen tension in the bubble (as it affects the oxygen deficit), and the travel time of the bubble in the water column. The air–water interface may be increased by decreasing the size of the bubble, and the oxygen concentration gradient between the bubble and the water can be increased by using pure oxygen or increasing the pressure of the injected gas.

Colt and Tchobanoglous (1981) list oxygen transfer rates for diffused air systems under standard conditions at 0.6 to 2.0 kg/kWh, depending upon bubble size. The distance of travel time of the bubble in the water column is not given.

When correcting standard oxygen transfer rates for field conditions, Equation 8.3 may be used, but the Ce_t term must be corrected for the partial pressure of oxygen in the bubble. For instance, if pure oxygen is delivered at 1.0 atm of pressure,

$$Ce'_t = Ce_t \times \frac{760 \text{ mm Hg}}{159.2 \text{ mm Hg}} \qquad (8.4)$$

where Ce'_t = the solubility of oxygen with a pure oxygen atmosphere, 760 mm Hg = the partial pressure of oxygen in pure oxygen at a pressure of 1.0 atm, and 159.2 mm Hg = the partial pressure of oxygen in air at a pressure of 1.0 atm. The corrected Ce'_t term for diffused air systems using compressed air is

$$Ce'_t = Ce_t \times \frac{\text{pressure of compressed air}}{\text{atmospheric pressure}} \qquad (8.5)$$

The air pressure from the compressor may be given in pounds per square inch (psi). The conversion from psi to mm Hg is 0.0193 psi/mm Hg.

Oxygenation

Most modern flowing water aquaculture applications have abandoned mechanical aeration in favor of using liquid oxygen (LOX). Pure oxygen can be produced on-site, but is more commonly delivered by truck and stored in vessels that are purchased or leased (Figure 8.10).

If bulk liquid oxygen is not available, it can be produced using a device called a pressure swing adsorption (PSA) unit. The PSA unit forces compressed air through a molecular sieve that adsorbs nitrogen, producing a gas that is 85% to 95% oxygen (Boyd and Watten 1989). A PSA system capable of generating 17 m³/hr (25 kg/hr) costs $30,000 (Boerson and Chessney 1986). Operation of this unit would require a 40 HP (29.8 kW) compressor and backup electrical generator that would cost an additional $25,000. A smaller unit producing 2.4 m³/hr (3kg/hr) costs $5,850 and the required 9.4 HP (7.0 kW) compressor could be purchased for $8,100 (Watten 1991).

Figure 8.10 Rented tank to store liquid oxygen for aquaculture.

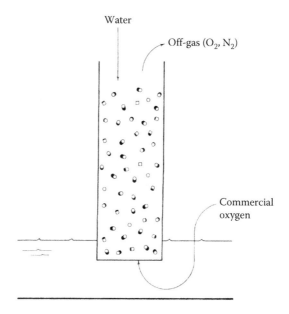

Figure 8.11 Enclosed packed column for absorption of oxygen. (From Speece, R. E. 1981. *Proceedings of the Bioengineering Symposium for Fish Culture*, Allen, L. J. and Kinney, E. C., Eds., American Fisheries Society, Bethesda, Maryland. With permission.)

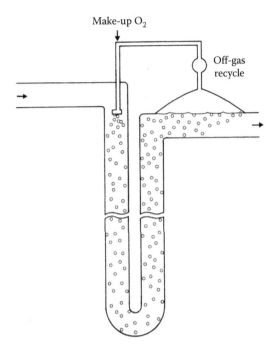

Figure 8.12 U-tube with off-gas recycle for absorption of oxygen. (From Speece, R. E. 1981. Management of dissolved oxygen and nitrogen in fish hatchery waters. Pages 53–62 *in* L. J. Allen and E. C. Kinney, editors. *Proceedings of the Bioengineering Symposium for Fish Culture.* Fish Culture Section Special Publication 1, American Fisheries Society, Bethesda, Maryland. With permission.)

Speece (1981) described five devices capable of transferring 90% or more of diffused oxygen to water under typical aquaculture conditions. They are enclosed packed columns (Figure 8.11), U-tube (Figure 8.12), downflow bubble contact aerator (Figure 8.13), recycled diffused oxygenation (Figure 8.14), and rotating packed column (Figure 8.15) devices.

The dimensions and operation of a packed column (Figure 8.11) are given by Owsley (1981). A 0.25 m diameter tube, 1.52 m tall, packed with 3.8 cm diameter plastic rings (see Figure 11.2) was supplied with a water flow of 568 L/min. When oxygen was injected at the bottom of the column so that it rose counter-current to the water flow, the effluent DO concentration was in excess of 40 mg/L when influent DO was 6.0 mg/L. An oxygen absorption rate of 90% was achieved.

A U-tube (Figure 8.12) is a deep hole with a baffle in the center that forces water to the bottom then back up to the surface. Gas is injected into the inlet side of the U-tube, and the pressure increase as the bubbles descend the tube enhances gas transfer. Speece (1981) reported that with a tube depth of 12.2 m, water velocity of 1.8 m/sec, and influent DO of

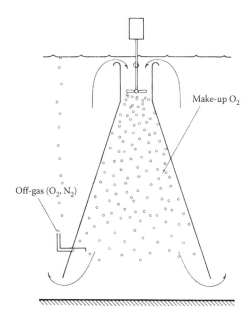

Figure 8.13 Downflow bubble contact aerator for absorption of oxygen. (From Speece, R. E. 1981. Management of dissolved oxygen and nitrogen in fish hatchery waters. Pages 53–62 *in* L. J. Allen and E. C. Kinney, editors. *Proceedings of the Bioengineering Symposium for Fish Culture*. Fish Culture Section Special Publication 1, American Fisheries Society, Bethesda, Maryland. With permission.)

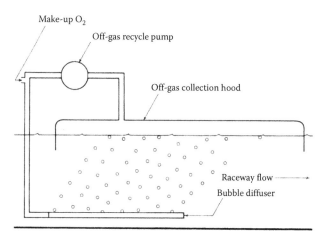

Figure 8.14 Recycled diffused oxygenation for absorption of oxygen in shallow ponds. (From Speece, R. E. 1981. Management of dissolved oxygen and nitrogen in fish hatchery waters. Pages 53–62 *in* L. J. Allen and E. C. Kinney, editors. *Proceedings of the Bioengineering Symposium for Fish Culture*. American Fisheries Society, Bethesda, Maryland. With permission.)

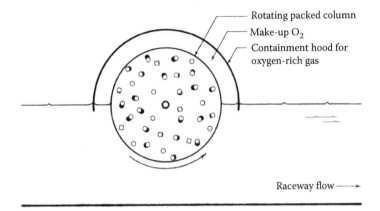

Figure 8.15 Rotating packed column for absorption of oxygen in shallow ponds. (From Speece, R. E. 1981. Management of dissolved oxygen and nitrogen in fish hatchery waters. Pages 53–62 *in* L. J. Allen and E. C. Kinney, editors. *Proceedings of the Bioengineering Symposium for Fish Culture.* Fish Culture Section Special Publication 1, American Fisheries Society, Bethesda, Maryland. With permission.)

6.0 mg/L, effluent DO would be about 40 mg/L and 90% of injected gas would be absorbed if a recycle hood is located at the effluent end to trap off-gas and recycle it back down the tube. The head loss through the unit under these flow conditions was 1.2 m.

The downflow bubble contact aerator (Figure 8.13) consists of a surface agitator at the top of a 3m high cone. Oxygen is injected into the cone and the bubbles are trapped inside because the water velocity is greater at the top of the cone than at the bottom. Thus, the contact time of the bubbles in the water is greatly increased. Discharged DO from this device should be 22 mg/L when influent DO is 6.0 mg/L and 90% of the injected oxygen is absorbed (Speece 1981).

Speece (1981) also described two devices that can be placed directly into shallow raceways to achieve 90% oxygen absorption. Recycled diffused oxygenation (Figure 8.14) achieves this by continuously recycling off-gas back to the diffuser with a small blower. The rotating packed column (Figure 8.15) is a tube filled with plastic rings (see Figure 11.2), half submerged in the water column and rotating counter-current to the raceway flow. The unit is enclosed in a hood that contains a pure oxygen atmosphere.

The multi-stage column system (Watten 1991) or a downflow bubble contactor is almost always selected for aquaculture applications. Watten (1991) described a multi-stage column system he called the low head oxygenator (LHO) in which the head requirement for oxygenation is reduced by employing a series of short columns. The column bank is operated in

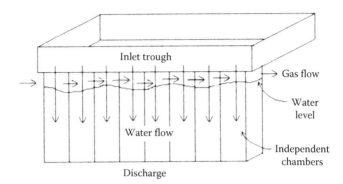

Figure 8.16 Low head oxygenator (LHO) in which water flows parallel and oxygen flows in series through a bank of short columns. (From Watten, B. J. 1991. Application of pure oxygen in raceway culture systems. Pages 311–332 *in Engineering Aspects of Intensive Aquaculture*. Northeast Regional Aquaculture Engineering Service, Ithaca, New York. With permission.)

parallel to water flow and in series with gas flow (Figure 8.16 and 8.17). Thus, off-gas is repeatedly contacted with inlet water and a high gas transfer is achieved through distances of waterfall of 0.3 to 1.0 m. Watten's (1991) invention of the LHO has resulted in the widespread adoption of liquid oxygen for the aquaculture of salmonids. The LHO fits conveniently into a raceway (Figure 8.18).

If dual-drain circular tanks are selected for serial reuse, the LHO could be placed in a sump between rearing units, but the best choice in

Figure 8.17 LHO with top plate removed showing detail of chambers.

Figure 8.18 The LHO fits conveniently into a raceway.

this case is a downflow bubble contactor. The downflow bubble contactor commonly used to oxygenate circular tanks differs from that described by Speece (1981) in that it does not require a propeller and is placed directly in a pipeline, thus eliminating the need for a sump (Figure 8.19).

Water enters the top of the cone and oxygen is introduced to the bottom. The cone shape allows for efficient oxygen transfer by reducing water velocity to less than the rise velocity of the oxygen bubbles (Losordo 1997). Schematic diagrams of both devices can be found in Watten (1991) and Losordo et al. (1999).

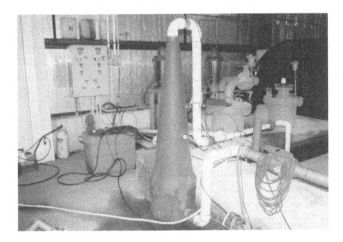

Figure 8.19 Downflow bubble contactor used to add liquid oxygen to a pipeline.

An important feature of pure oxygen is that DO levels can easily be maintained at levels above saturation. This allows for carrying capacity to be increased over that possible with water with DO at atmospheric equilibrium. In practice, oxygen flow to the LHO or downflow bubble contactor is adjusted so that rearing unit effluent DO levels do not decrease below the designated minimum allowable concentration. Since LOX is relatively inexpensive and easily transferred to water, target effluent DO levels should be 100 mm Hg to ensure optimum fish health. The current cost of liquid oxygen (in 2016), delivered within 100 miles from the air reduction plant, is approximately $0.57/100 ft^3 ($ 0.109/kg). Rentals for 1500 gallon (8978 kg) and 3000 gallon (17,956 kg) on-site storage tanks are $500/month and $700/month, respectively.

Problems with gas supersaturation

Occasionally, groundwater supplies used for aquaculture are supersaturated with nitrogen gas. This condition is often associated with undersaturation of oxygen. Supersaturation of dissolved nitrogen (DN) can only be accomplished by a change in temperature or pressure that lowers the solubility of the gases in water. Groundwater becomes supersaturated with dissolved gases when cool winter rains and snowmelt enter a warmer aquifer. Dissolved oxygen levels may be reduced by microbial respiration in the soil. Water may become supersaturated with atmospheric gases by a faulty pump that draws air in at the suction end or by heating water to accelerate fish growth. The changes in gas saturation associated with temperature changes may be expressed as

$$\text{Percent Saturation} = Ce_{t1}/Ce_{t2} \qquad (8.6)$$

where Ce_{t1} = the gas concentration at temperature 1 and Ce_{t2} = the equilibrium concentration of that gas at temperature 2. Equilibrium concentrations of DN are presented in Table 8.2. Suppose water saturated with atmospheric gases at 10°C is heated to 20°C. The resulting saturation of DO is 10.92/8.84 = 123% and of DN is 18.14/14.88 = 122%.

When fish are exposed to supersaturated levels of DN they may experience a condition known as gas bubble disease, in which nitrogen bubbles form in the blood and block capillaries by embolism. There is considerable evidence that only the inert gases dissolved in water cause gas bubble disease. Exposure to supersaturated levels of DO or CO_2, which commonly occurs in fish transport and in static water aquacultures with high rates of photosynthesis and respiration, does not cause gas embolism; apparently bubbles that form are metabolized before embolism occurs.

Table 8.2 Equilibrium concentrations of dissolved nitrogen in mg/L at T°C
and 1.0 atm of pressure

T	0.0	0.1	0.2	0.3	0.4	0.5	0.6	0.7	0.8	0.9
4	20.82	20.77	20.72	20.67	20.62	20.57	20.52	20.47	20.42	20.38
5	20.33	20.28	20.23	20.18	20.13	20.09	20.04	19.99	19.85	19.90
6	19.85	19.81	19.76	19.71	19.67	19.62	19.56	19.53	19.49	19.44
7	19.40	19.35	19.31	19.26	19.22	19.18	19.13	19.09	19.05	19.00
8	18.96	18.92	18.88	18.83	18.79	18.75	18.71	18.67	18.63	18.58
9	18.54	18.50	18.46	18.42	18.38	18.34	18.30	18.26	18.22	18.18
10	18.14	18.10	18.06	18.02	17.99	17.95	17.91	17.87	17.83	17.79
11	17.76	17.72	17.68	17.64	17.61	17.57	17.53	17.50	17.46	17.42
12	17.39	17.35	17.31	17.28	17.24	17.21	17.17	17.13	17.10	17.06
13	17.03	16.99	16.96	16.93	16.89	16.86	16.82	16.79	16.75	16.72
14	16.69	16.65	16.62	16.59	16.55	16.52	16.49	16.45	16.42	16.39
15	16.36	16.32	16.29	16.26	16.23	16.20	16.16	16.13	16.10	16.07
16	16.04	16.01	15.98	15.96	15.91	15.88	15.85	15.82	15.79	15.76
17	15.73	15.70	15.67	15.64	15.61	15.58	15.55	15.52	15.50	15.47
18	15.44	15.41	15.38	15.35	15.32	15.29	15.27	15.24	15.21	15.18
19	15.15	15.12	15.10	15.07	15.04	15.01	14.99	14.96	14.93	14.91
20	14.88	14.85	14.82	14.80	14.77	14.74	14.72	14.69	14.67	14.64
21	14.61	14.59	14.56	14.53	14.51	14.48	14.46	14.43	14.41	14.38
22	14.36	14.33	14.31	14.28	14.26	14.23	14.21	14.18	14.16	14.13
23	14.11	14.08	14.06	14.04	14.01	13.99	13.96	13.94	13.92	13.89
24	13.87	13.85	13.82	13.80	13.78	13.75	13.73	13.71	13.68	13.66
25	13.64	13.61	13.59	13.57	13.55	13.52	13.50	13.48	13.46	13.44
26	13.41	13.39	13.37	13.35	13.33	13.30	13.28	13.26	13.24	13.22
27	13.20	13.17	13.15	13.13	13.11	13.09	13.07	13.05	13.03	13.01
28	12.99	12.96	12.94	12.92	12.90	12.88	12.86	12.84	12.82	12.80
29	12.78	12.76	12.74	12.72	12.70	12.68	12.66	12.64	12.62	12.60
30	12.58	12.50	12.54	12.53	12.51	12.49	12.47	12.45	12.43	12.41

Source: Data from Colt, J. E. 1984. *Computation of Dissolved Gas Concentrations in Water as Functions of Temperature, Salinity, and Pressure.* American Fisheries Society Special Publication 14. Bethesda, Maryland.

Bubbles can occur in fish blood only when the gas pressure in water exceeds the hydrostatic pressure. A 10m submergence depth produces a hydraulic pressure of approximately 1 atm. For example, if the gas pressure in water is 890 mm Hg and the barometric pressure is 760 mm Hg, the gas saturation of the water is 890/760 = 117%. Bubbles can form if ΔP, the difference between total gas pressure and hydrostatic pressure, is positive. If the fish is at a depth of 1 m, the hydrostatic pressure is 76 mm Hg and ΔP = 890–(760 + 76) = 54 mm Hg. Therefore, bubbles could form. If the fish submerges to a depth of 2 m, $\Delta P = 890 - (760 + 152) = -22$ mm Hg, bubbles could not form. Unfortunately, fish in culture do not have the luxury of choosing a favorable depth and gas bubble disease can cause serious losses at hatcheries whose water supplies are supersaturated with DN.

The data on toxic levels of DN are very conflicting (Weitkamp and Katz 1980). Fish eggs are probably not affected by nitrogen supersaturation. Meekin and Turner (1974) reported mortality in steelhead eggs incubated in water having a total gas pressure of 112% saturation, but factors other than gas pressure may have caused this mortality (Weitkamp and Katz 1980). In general, tolerance to supersaturation increases with fish age, with the most serious losses to fry and juveniles. Salmonids appear to be less tolerant of gas supersaturation than most other species. The critical level of DN saturation is generally considered to be 110% saturation (Thurston et al. 1979).

Because supersaturation of water is an unstable condition that tends to return to equilibrium, aeration, or, in this case, deaeration, is a logical remedy. Due to their relative solubilities, nitrogen is 1.51 times easier to transfer to water than oxygen (Colt and Westers 1982). Thus, the mass transfer equation for DN removal may be written as

$$RT_{DN} = 1.51 \, RS_{DO} \frac{(C - Ce)(1.025^{T-20})(0.85)}{14.88} \tag{8.7}$$

where $RT_{DN} = $ DN removal at conditions at the aquaculture site, $RS_{DO} = $ the oxygen transfer for a particular aerator at standard conditions, $C = $ the DN concentration in mg/L at the aquaculture site, $Ce = $ the equilibrium concentration of DN, and $T = $ the water temperature in °C. The equilibrium concentration of DN at standard conditions is 14.88 mg/L (Table 8.2).

Suppose a mechanical aerator that transfers oxygen at 2 kg/kWh at standard conditions is used to remove excess DN from water whose temperature is 10°C, total gas pressure is 115% saturation, and barometric pressure is 740 mm Hg. Then $RS_{DO} = 2$ kg/kWh, $Ce = 18.14$ (740/760) = 17.66 mg/L, and $C = 17.66 \times 115\% = 20.31$ mg/L. Thus,

$$RT_{DN} = (1.51)(2.0)\frac{(20.31 - 17.66)(1.025^{-10})(0.85)}{14.88} = 0.36 \text{ kgDN/kWh}$$

Note that the actual nitrogen transfer from water to air is very low because the driving force $(C - Ce)$ is small. Now, suppose that a flow of 2000 L/min is to be degassed with a desired effluent level of 100% saturation. The amount of DN that must be removed is $20.31 - 19.43 = 0.88$ mg/L. Then,

$$\frac{0.88 \text{ mg}}{\text{L}} \times \frac{2000 \text{ L}}{\text{min}} \times \frac{\text{kg}}{10^6 \text{ mg}} \times \frac{\text{kWh}}{0.36 \text{ kg}} \times \frac{60 \text{ min}}{\text{hr}} = 0.29 \text{ kW}$$

and an aerator of at least 0.29 kW is required.

Packed columns have been used successfully to remove excess DN from hatchery water supplies. Owsley (1981) reported that a 0.25 m diameter tube, 1.52 m high, filled with 3.8-cm plastic rings, supplied with 379 to 568 L/min of water, reduced DN levels from 130% to near 100% saturation.

Speece (1981) reported that any device that successfully adds pure oxygen to water will result in DN levels of less than 100% saturation because oxygen displaces the excess nitrogen.

SAMPLE PROBLEMS

1. A hatchery with 12°C water at an elevation of 600 m above sea level has raceways arranged in a 3-unit series with a 30-cm drop between ponds. The series receives 1800 L/min of water. The hatchery manager is requested to produce 25-cm brown trout. How many fish should be stocked in each pond of the series if a simple weir separates them and influent DO to the system is at 100% saturation? If lattice gravity aerators are installed, how much could stocking rates be increased? If the fish are valued at $5.00/kg, what is the economic benefit of installing the lattices?

2. A 1.5 kW mechanical aerator is rated by the manufacturer to deliver oxygen at 1.5 kg/kWh. If this unit is placed in a raceway receiving 12,000 L/min of 10°C water at 150 m above sea level, at a point where fish respiration has decreased the DO to 5.0 mg/L, how much will the DO be increased below the unit? If a second aerator is installed below the first, how much will the DO be increased below it?

3. How large of a mechanical aerator (kW) should be selected for the hatchery described in Problem 2?

4. How much LOX would be required to properly oxygenate the water supply in Problem 2?

5. Which mode of aeration would be the most economical for this example if electric power is valued at $0.15/kWh?

6. A well delivers 2000 L/min of 9°C water at an elevation of 300 m above sea level. The DO concentration is 25% of saturation and the DN level is 125% of saturation. What size mechanical aerator is required to achieve acceptable dissolved gas levels for fish culture?

7. Estimate the costs for mechanical aeration versus LOX delivered through an LHO or downflow bubble contactor to achieve acceptable dissolved gas levels for the example in Problem 6.

References

Boerson, G. and J. Chessney. 1986. Engineering considerations in supplemental oxygen. *Northwest Fish Culture Conference*, Springfield, Oregon.

Boyd, C. E. and B. J. Watten. 1989. Aeration systems in aquaculture. *CRC Critical Reviews in Aquatic Sciences* 1: 425–472.

Burrows, R. E. and B. D. Combs. 1968. Controlled environments for salmon propagation. *Progressive Fish-Culturist* 30: 123–136.

Chesness, J. L. and J. L. Stephens. 1971. A model study of gravity flow cascade aerators for catfish raceway systems. *Transactions of the American Society of Agricultural Engineers* 14: 1167–1169, 1174.

Colt, J. E. 1984. *Computation of Dissolved Gas Concentrations in Water as Functions of Temperature, Salinity, and Pressure*. American Fisheries Society Special Publication 14. Bethesda, Maryland.

Colt, J. E. and G. Tchobanoglous. 1981. Design of aeration systems for aquaculture. Pages 138–148 *in* L. J. Allen and E. C. Kinney, editors. *Proceedings of the Bio-Engineering Symposium for Fish Culture*. Fish Culture Section Publication 1, American Fisheries Society, Bethesda, Maryland.

Colt, J. E. and H. Westers. 1982. Production of gas supersaturation by aeration. *Transactions of the American Fisheries Society* 111: 342–360.

Downing, A. L. and G. A. Truesdale. 1955. Some factors affecting the rate of solution of oxygen in water. *Journal of Applied Chemistry* 5: 570–581.

Eckenfelder, W. W., Jr. 1970. Oxygen transfer and aeration. Pages 1–12 *in* W. W. Eckenfelder, editor. *Manual of Treatment Processes*, Vol. 1. Water Resources Management Series, Environmental Sciences Service Corporation, Stamford, Connecticut.

Haskell, D. C., R. O. Davies, and J. Reckahn. 1960. Factors affecting hatchery pond design. *New York Fish and Game Journal* 7: 113–129.

Losordo, T. M. 1997. Tilapia culture in intensive recirculating systems. Pages 185–211 *in* B. A. Costa-Pierce and J. E. Rakocy, editors. *Tilapia Aquaculture in the Americas*, Vol. 1. The World Aquaculture Society, Baton Rouge, Louisiana.

Losordo, T. M., M. B. Masser, and J. E. Rakocy. 1999. *Recirculating Aquaculture Tank Production Systems: A Review of Component Options*. Southern Regional Aquaculture Center, Publication No. 453. College Station, Texas.

Mayo, R. D. 1979. A technical and economic review of the use of reconditioned water in aquaculture. Pages 508–520 *in* T. V. R. Pillay and W. A. Dill, editors. *Advances in Aquaculture*. FAO Technical Conference on Aquaculture, Kyoto, Japan.

Meekin, T. K. and B. K. Turner. 1974. Tolerance of salmonid eggs, juveniles and squawfish to supersaturated nitrogen. *Washington Department of Fisheries Technical Report* 12: 78–126.

Owsley, D. E. 1981. Nitrogen gas removal using packed columns. Pages 71–82 *in* L. J. Allen and E. C. Kinney, editors. *Proceedings of the Bio-Engineering Symposium for Fish Culture*. Fish Culture Section Publication 1, American Fisheries Society, Bethesda, Maryland.

Soderberg, R. W. 1982. Aeration of water supplies for fish culture in flowing water. *Progressive Fish-Culturist* 44: 89–93.

Soderberg, R. W., J. B. Flynn, and H. R. Schmittou. 1983. Effects of ammonia on growth and survival of rainbow trout in intensive static-water culture. *Transactions of the American Fisheries Society* 112: 448–451.

Speece, R. E. 1981. Management of dissolved oxygen and nitrogen in fish hatchery waters. Pages 53–62 *in* L. J. Allen and E. C. Kinney, editors. *Proceedings of the Bio-Engineering Symposium for Fish Culture.* Fish Culture Section Special Publication 1, American Fisheries Society, Bethesda, Maryland.

Tebbutt, T. H. Y. 1972. Some studies on reaeration in cascades. *Water Research* 6: 297–304.

Thurston, R. V., R. C. Russo, C. M. Feterolf Jr., T. A. Edsall, and Y. M. Barber Jr., editors. 1979. *A Review of the EPA Red Book: Quality Criteria for Water.* Water Quality Section, American Fisheries Society, Bethesda, Maryland.

Watten, B. J. 1991. Application of pure oxygen in raceway culture systems. Pages 311–332 *in* Engineering Aspects of Intensive Aquaculture. Northeast Regional Aquacultural Engineering Service, Cornell University, Ithaca, New York.

Weitkamp, D. E. and M. Katz. 1980. A review of dissolved gas supersaturation literature. *Transactions of the American Fisheries Society* 109: 659–702.

Westers, H. and K. M. Pratt. 1977. Rational design of hatcheries for intensive salmonid culture, based on metabolic characteristics. *Progressive Fish-Culturist* 39: 157–165.

Whipple, W., Jr., J. V. Hunter, B. Davidson, F. Dittman, and S. Yu. 1969. *Instream Aeration of Polluted Rivers.* Water Resources Institute, Rutgers University, New Brunswick, New Jersey.

chapter nine

Ammonia production and toxicity

Ammonia production rate

Ammonia is the principle nitrogenous by-product of fish metabolism and is of importance in fish culture because it is toxic to fish in its un-ionized form. The origin of metabolic ammonia is the deamination of amino acids utilized as energy. A metabolic nitrogen budget allows for the estimation of the contribution of dietary protein to the accumulation of ammonia in water. For every 100 g of protein fed, approximately 40 g will be assimilated as fish flesh. If 20 g are undigested and 5 g are uneaten, 35 g will be metabolized as energy (Lovell 1989). Protein is approximately 16% nitrogen. Thus, fish consumption of 100 g of dietary protein will result in 35 g × 16%, or 5.6 g of ammonia being excreted. The ammonia production rate may be expressed as

$$A = 56P \tag{9.1}$$

where A = the ammonia production rate in g total ammonia nitrogen (TAN) per kg of food and P = the decimal fraction of protein in the diet. For example, feeding a 45% protein diet would result in the production of 25.2 g of ammonia nitrogen per kg of food fed. In this book, unless stated otherwise, concentrations of all nitrogenous species are expressed as mg/L of nitrogen, which is consistent with most of the literature.

Ammonia toxicity

Aqueous ammonia occurs in two molecular forms and the equilibrium between them is determined by pH and, to a lesser extent, temperature:

$$NH_3 \leftrightarrow NH_4^+$$

and

$$NH_3 - N + NH_4^+ - N = TAN$$

                                                                 Aquaculture technology

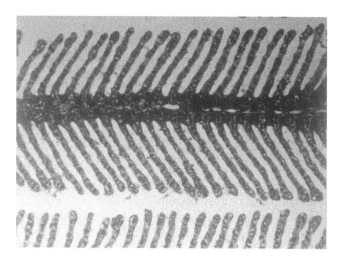

Figure 9.1 Normal rainbow trout gill magnified 75×. (From Soderberg, R. W. 1985. *Journal of Fish Diseases* 8: 57–64. With permission.)

The un-ionized form, NH_3, is a gas and can freely pass the gill membrane. The rate and direction of passage depends upon the NH_3 concentration gradient between the fish's blood and the adjacent water. Un-ionized ammonia is toxic to fish, whereas NH_4^+ is relatively nontoxic.

Chronic exposure to NH_3 damages fish gills, reducing the epithelial surface area available for gas exchange (Burrows 1964; Flis 1968; Bullock 1972; Larmoyeaux and Piper 1973; Smart 1976; Burkhalter and Kaya 1977; Thurston et al. 1978; Soderberg et al. 1984a,b; Soderberg 1985) (Figures 9.1 through 9.3). Snieszko and Hoffman (1963), Burrows (1964), Larmoyeaux and Piper (1973), Walters and Plumb (1980) and Soderberg et al. (1983) have reported that NH_3 exposure predisposes fish to disease. Reduction in fish growth caused by NH_3 exposure is widely documented (Brockway 1950; Kawamoto 1961; Burrows 1964; Smith and Piper 1975; Robinette 1976; Burkhalter and Kaya 1977; Colt and Tchobanoglous 1978; Soderberg et al. 1983). Gill damage caused by ammonia exposure may contribute to reduced growth by reducing oxygen consumption (Burrows 1964; Smith and Piper 1975) or the osmoregulatory cost of hyperventilation induced by reduced epithelial surface available for oxygen uptake (Lloyd and Orr 1969).

Calculation of ammonia concentration

The average daily TAN concentration in a fish rearing unit is the total daily ammonia production in mg divided by the total daily flow in L. For example, suppose that 1000 kg of fish are held in a pond receiving 500 L/ min of water and are fed 1.0% of their body weight per day of 40% protein

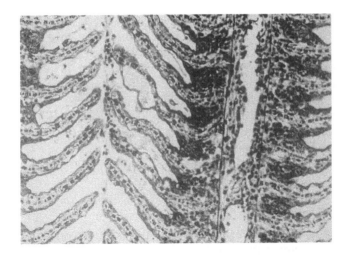

Figure 9.2 Rainbow trout gill with epithelial edema and hyperplasia at the bases of the filiments. These lesions are characteristic of ammonia exposure. Magnification is 150×. (From Soderberg, R. W. 1985. *Journal of Fish Diseases* 8: 57–64. With permission.)

food. The ammonia production rate is A = 56P = 56 (0.40) = 22.4 g/kg of food. The daily food ration is 1000 kg × 1.0% = 10 kg, and the daily ammonia production is 22.4 g/kg × 10 kg = 224 g (224,000 mg). The total daily water flow is 500 L/min × 1440 min/day = 720,000 L and the average daily TAN concentration is 224,000/720,000 = 0.31 mg/L.

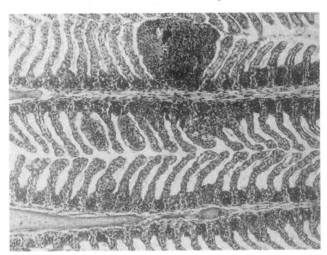

Figure 9.3 Rainbow trout gill with several large blood-filled aneurisms and hyperplasia causing fusion of some filaments. These lesions are characteristic of ammonia exposure.

Analytical procedures do not differentiate between the two forms of ammonia in solution, and only one form is of consequence to the fish culturist. Thus, it is important to be readily able to determine the fraction of NH_3 in a solution at any temperature and pH. Then, if the level of NH_3 above which toxic effects occur is known, waters may be characterized according to their reusability with respect to ammonia.

Emerson et al. (1975) present the following formula to calculate the acid dissociation constant, expressed as the negative log, for ammonia, based on the values of Bates and Pinching (1949):

$$pKa = 0.09018 + \frac{2729.92}{T + 273.15} \tag{9.2}$$

where pKa is the negative log of the acid dissociation constant for ammonia and T the temperature in °C.

A formula to solve for the NH_3 fraction, $NH_3/(NH_3 + NH_4^+)$, in an ammonia solution, is derived as follows. The acid dissociation for ammonia may be expressed as $NH_4^+ \leftrightarrow NH_3 + H^+$, and the equilibrium constant, Ka, is

$$Ka = \frac{(NH_3)(H^+)}{(NH_4^+)} = 10^{pKa}$$

Solving for the fraction $NH_3/NH_3 + NH_4^+) = f$,

$$f = \frac{1}{10^{pKa-pH} + 1} \tag{9.3}$$

The un-ionized fraction, f, is the decimal fraction of NH_3 in an ammonia solution. Thus, NH_3–N = TAN × f.

The use of these equations to calculate the un-ionized fraction is best illustrated by example. Suppose we wish to calculate f for 10°C water at pH values of 6, 7, and 8. From Equation 9.2,

$$pKa = 0.09018 + \frac{2729.92}{10 + 273.15} = 9.731$$

Substituting this into Equation 9.3 for a pH of 6,

$$f = \frac{1}{10^{9.731-6} + 1} = 0.000186$$

$$f \text{ at pH } 7 = \frac{1}{10^{9.731-7}+1} = 0.00185$$

$$f \text{ at pH } 8 = \frac{1}{10^{9.731-8}+1} = 0.0182$$

Note that an increase in pH of one unit increases the un-ionized fraction of ammonia approximately 10-fold. Thus, for a given total ammonia level, fish in low pH water will be exposed to less NH_3 than those in water with a higher pH. Because of its effect on the toxicity of ammonia, pH is one of the most important considerations in the selection of a water supply for intensive aquaculture.

Correction of the NH_3 fraction for ionic strength

Ionic strength, in addition to pH and temperature, controls the ionization of ammonia and can significantly affect the un-ionized fraction of ammonia in brackish or marine waters. The apparent toxicity of NH_3 in saline waters is reduced because, as ionic concentrations increase, electrostatic forces among ions and molecules increase the extent to which ionization occurs. Soderberg and Meade (1991) present the following procedure for correction of the NH_3 fraction in an ammonia solution for ionic strength. The strength of the electrostatic field that impinges on ionic activity is affected by the concentration and charge of ions in solution, and is expressed as ionic strength (I):

$$I = \sum \frac{(M_i)(z_i)^2}{2} \tag{9.4}$$

where M is the molecular concentration of a given ion (i) and z is its charge. For example, the ionic strength of a solution containing 100 mg/L each of Na^+ and Cl^- would be calculated as follows: Atomic weights are 23 for sodium and 35.5 for chloride. Note that Na^+ and Cl^- are both monovalent (single charge) ions.

$$\frac{100 \text{ mg Na}}{L} \times \frac{\text{mole}}{23,000 \text{ mg}} = 0.0043 \text{ M Na}$$

$$\frac{100 \text{ mg Cl}}{L} \times \frac{\text{mole}}{35,500 \text{ mg}} = 0.0028 \text{ M Cl}$$

$$I = \frac{(0.0043)(1)^2}{2} + \frac{(0.0028)(1)^2}{2} = 0.0036 \text{ M}$$

To calculate the un-ionized fraction of ammonia in the above solution, the salinity correction term, s, is calculated from

$$s = -\frac{A'\sqrt{I}}{1+\sqrt{I}} \tag{9.5}$$

Values for the coefficient A' are provided in Table 9.1. For the present example, if the pH is 7.5 and the temperature is 10°C,

$$s = -\frac{0.498\sqrt{0.0036}}{1+\sqrt{0.0036}} = -0.028$$

Table 9.1 Values for the coefficient A' used to correct ammonia ionization for ionic strength

T (°C)	A'	T (°C)	A'
0	0.492	18	0.506
1	0.492	19	0.506
2	0.493	20	0.507
3	0.494	21	0.508
4	0.495	22	0.509
5	0.495	23	0.510
6	0.496	24	0.511
7	0.497	25	0.512
8	0.497	26	0.513
9	0.498	27	0.514
10	0.498	28	0.515
11	0.500	29	0.515
12	0.501	30	0.516
13	0.501	31	0.517
14	0.502	32	0.518
15	0.503	33	0.519
16	0.504	34	0.520
17	0.505	35	0.521

Source: Data from Soderberg, R. W. and J. W. Meade. 1991. *Progressive Fish-Culturist* 53: 118–120.

which is inserted into Equation 9.3 as follows:

$$f = \frac{1}{10^{pKa-pH-s}+1}$$

and

$$f = \frac{1}{10^{9.731-7.5+0.028}+1} = 0.00547$$

If f is not corrected for ionic strength,

$$f = \frac{1}{10^{9.731-7.5}+1} = 0.00584$$

and the un-ionized fraction is overestimated by 6.3%.

Assignment of NH_3 maxima for hatcheries

The maximum level of NH_3 at which toxic effects are tolerable may be selected as a hatchery design criterion in a manner similar to that used to assign dissolved oxygen (DO) minima (Chapter 6). Robinette (1976) reported reduced growth in catfish at NH_3 concentrations above 0.12 mg/L. Colt and Tchobanoglous (1978) found that catfish growth was reduced at all concentrations above 0.048 mg/L, and that the growth rate at 0.517 mg/L was half that at 0.048 mg/L. Larmoyeaux and Piper (1973) reported reduced growth of rainbow trout at 0.0166 mg/L, but not at 0.0125 mg/L. The U.S. Environmental Protection Agency (Thurston et al. 1979) has established an acceptable level of NH_3 for aquatic organisms of 0.016 mg/L. This is probably a reasonable design criterion for intensive aquaculture although catfish, and possibly other species, are apparently less sensitive to NH_3 than are trout, and higher NH_3 maxima may be acceptable.

Muir et al. (2000) recommended that NH_3 levels be kept below 1 mg/L and, ideally, below 0.2 mg/L for flowing water tilapia culture, but there are no bioassay data available on the chronic toxicity of NH_3 to tilapia. Ball (1967) reported that the 2-day LC_{50} (concentration resulting in 50% mortality) of NH_3 to rainbow trout is 0.41 mg/L. Colt and Tchobanoglous (1978) reported the LC_{50} of NH_3 to channel catfish is 2–3.1 mg/L. Rainbow trout and channel catfish exhibit reduced growth at NH_3 concentrations of 0.016 mg/L (Larmoyeaux and Piper 1973) and 0.048 mg/L (Colt and Tchobanoglous 1978), respectively. Thus, the application factors between chronic and acute toxicities of NH_3 are 0.04 (0.016/0.041) for rainbow trout and 0.015–0.024 for catfish. The 72-hour LC_{50} of NH_3 to blue tilapia (*O. aureus*) has been reported at 2.35 mg/L (Redner and Stickney 1979). Applying application factors of 0.015 and 0.04 to this value results in projected chronic toxicities of 0.035 to 0.094 mg/L for tilapia. More research is required to define the maximum allowable exposures of NH_3 for tilapia because this is a critical parameter in designing flowing water aquaculture systems.

Carrying capacity with respect to ammonia

Incorporation of ammonia toxicity considerations into hatchery design where liquid oxygen is not employed is best explained in terms of permissible water uses before NH_3 accumulates to the designated maximum level. Water use is defined as the carrying capacity, with respect to DO, of reaerated water. Consider the following hatchery data: Ce = 10 mg/L, reaerated influent DO = 90% of saturation (9.0 mg/L), minimum allowable DO = 75 mm Hg (4.9 mg/L), F = 1.2% body weight per day, P = 0.42, pH = 7.7, T = 12°C, and maximum allowable NH_3 = 0.016 mg/L. Carrying capacity with respect to DO is calculated from Equation 7.1,

$$CC = \frac{(9.0 - 4.9)(0.0545)}{0.012} = 18.62 \text{ lb/gpm} = 2.24 \text{ kg/L min}$$

The TAN concentration after one water use is

$$2.24 \text{ kg fish} \times \frac{0.012 \text{ kg food}}{\text{kg fish}} \times \frac{56(0.42)\text{g TAN}}{\text{kg food}}$$
$$\times \frac{1000 \text{ mg}}{\text{g}} \times \frac{\text{L}}{\text{min}} \times \frac{1440 \text{ min}}{\text{day}} = 0.439 \text{ mg TAN/L}$$

Then use Equation 9.1 to calculate the pKa and Equation 9.3 to calculate the un-ionized fraction,

$$pKa = 009018 + \frac{2729.92}{273.15 + 12} = 9.664$$

$$f = \frac{1}{10^{9.664-7.7} + 1} = 0.0108$$

The NH_3 is 0.439 × 0.0108 = 0.0047 mg/L and the number of permissible water uses is 0.016/0.0047 = 3.4. Thus, the carrying capacity with respect to ammonia exposure is 2.24 kg/L min × 3.4 = 7.62 kg/L min.

Liquid oxygen is used to satisfy the respiratory requirements of cultured fish at most modern salmonid hatcheries. Oxygen is usually supplied through a low head oxygenator (LHO) at the head of the raceway and the oxygen flow is regulated so that effluent DO levels remain above 100 mm Hg. Thus, oxygen is not an issue and carrying capacity is determined by ammonia accumulation alone. Going back to the original example, 1.0 kg fish produces

$$\frac{0.012\,\text{kg food}}{\text{kg fish}} \times \frac{56\,(0.42)\,\text{g TAN}}{\text{kg food}} \times \frac{1000\,\text{mg}}{\text{g}} \times \frac{0.0108\,\text{mg NH}_3}{\text{mg TAN}} = 3.05\,\text{mg NH}_3$$

per day. If this is dissolved in 1 L/min (1440 L/day), the NH_3 concentration resulting from 1.0 kg fish/L/min is 0.00212 mg/L, and the fish load associated with a level of 0.016 mg/L NH_3 is 0.016/0.00212 = 7.55 kg/L min, but the fish in the liquid oxygen example would be exposed to a healthier DO regime than those in the aerated example.

Production capacity assessment

Recommended NH_3 maxima for fish hatcheries range from 0.0125 mg/L (Larmoyeaux and Piper 1973) to 0.016 mg/L (Willingham et al. 1979). Meade (1985) reviewed the literature on ammonia toxicity to fish and concluded that unknown site-specific water quality characteristics may significantly affect NH_3 toxicity and, thus, the maximum safe exposure level. Soderberg and Meade (1992) summarized water quality influences on the toxicity of NH_3 to fish. If the metabolite effects on cultured fish cannot be explained by NH_3 alone, a bioassay procedure technique such as the production capacity assessment (PCA) provided by Meade (1988) may be necessary for predicting hatchery carrying capacity.

The PCA procedure involves the experimental rearing of fish in a series of five or more containers so that fish in each rearing unit are exposed to the accumulated metabolites from the upstream units. Because this is a procedure to estimate carrying capacity independent of DO, it is essential that the water be reaerated to at least 90% of saturation at the influent of each rearing container. The water flow through the series is adjusted so that the fish in each container remove 25%–30% of the DO. Each container contains an equal weight of fish that are fed at the same rates. At the end of a 2- to 6-week period the oxygen consumption in each of the serial containers is determined by subtracting the effluent from influent levels before feeding. Fish are weighed and the specific growth rate of the fish in each container is regressed against cumulative oxygen consumption. The specific growth rate is log total fish weight at the end of the growth trial, minus log fish weight at the beginning of the growth trial, divided by the number of days in the growth trial, multiplied by 100. The cumulative oxygen consumption at which specific growth is reduced to a predetermined minimum, expressed as a percentage of the specific growth in the first container, represents carrying capacity in terms of oxygen use. Meade (1988) defined this level as the Effective Cumulative Oxygen Consumption (ECOC). The recommended minimum allowable specific growth reduction is 50% of the observed maximum during the PCA test, and the cumulative oxygen consumption that results in this level of growth reduction is defined as the $ECOC_{50}$ (Meade 1988). Thus, carrying capacity is defined as the weight of fish per unit of flow of water that extracts the $ECOC_{50}$ of DO from the water supply.

The following is an example of the calculations required for the PCA procedure. Calculation of the $ECOC_{50}$, based on the data in Table 9.2, is done in three steps, the results of which are shown in Table 9.3. The

Table 9.2 Fish weights and dissolved oxygen concentrations in 5 serial reuse units used in the example of the calculations in PCA

Serial unit	Weight (g)			DO (mg/L)	
(water use)	Starting	Ending	Cumulative	In	Out
1	4495	5386	5386	10.5	7.8
2	4505	5271	10,657	9.9	6.7
3	4475	4938	15,595	9.0	6.2
4	4372	4648	20,243	9.0	6.6
5	3861	3897	24,140	8.8	6.6

Source: Meade, J. W. 1988. *Aquacultural Engineering* 7: 139–146.

Note: The test was run for 15 days.

Table 9.3 Calculation of effective cumulative oxygen consumption (ECOC) below which 50% of maximum growth is achieved ($ECOC_{50}$), based on data in Table 9.1

Serial unit	Oxygen consumption		Specific growth rate
	Per unit	Cumulative	
1	2.7	2.7	0.561
2	3.2	5.9	0.488
3	2.8	8.7	0.305
4	2.4	11.1	0.190
5	2.2	13.3	0.029

determination of cumulative oxygen consumption through the series of five rearing units is shown in columns 2 and 3 of Table 9.3. The specific growth rate of fish in each container is shown in column 4 of Table 9.3. The final step is the determination, by linear regression, of the $ECOC_{50}$, in mg DO/L, which is completed by regressing the data in column 3 (Table 9.3) on that in column 4 (Table 9.3). Results follow:

Correlation coefficient $= -0.985$
Intercept $= 14.3$
Slope $= -19.0$
Specific growth rate for 50% of maximum growth $= 0.28$
$ECOC_{50} = -19.0 \, (0.28) + 14.3$
$ECOC_{50} = 9.0 \, mg/L$

Carrying capacity is estimated by regressing cumulative fish load (Table 9.2, column 4) on cumulative oxygen consumption (Table 9.3, column 3) and using the regression equation to calculate fish biomass whose oxygen consumption would equal 9.0 mg/L. The regression equation resulting from the data presented in this example is cumulative fish load $= 1781 \times$ cumulative oxygen consumption $+ 354$. When the $ECOC_{50}$ of 90 is substituted for cumulative oxygen consumption, the calculated cumulative fish load is 16,383 g. If the flow through the PCA rearing units is 3.0 L/min, the carrying capacity for the water supply would be 16.4 kg for 3.0 L/min, or 5.5 kg per L/min.

Soderberg (1995) predicted carrying capacities of three hatchery water supplies using the procedures here for estimating the biomass of 10 g lake trout that would result in a maximum NH_3 exposure of 0.016 mg/L. Then he determined the carrying capacities of these water supplies by conducting PCA bioassays using 10 g lake trout. The objective of the experiment was to compare the results from the two methods of calculating hatchery production capacity.

The water supplies tested were at Wellsboro (pH 7.0) and Lamar (pH 7.3), Pennsylvania, and Leetown, West Virginia (pH 7.7). Ammonia production per unit of the 42% protein diet used in the parallel PCA tests was estimated from the dietary nitrogen budget of Lovell (1989). The ammonia production rate per kg of fish was based on the feeding rate of Haskell (1959) and the growth rate determined from the linear equation from Soderberg (1992). The NH_3 fraction was calculated from Equation 9.3. Carrying capacity was calculated as kg of fish per L/min of flow that would result in an effluent NH_3 concentration of 0.016 mg/L.

Simultaneous PCA bioassays were conducted in triplicate at the three hatchery sites using the procedures devised by Meade (1988), with lake trout from a single lot of fish. Each of 5 bioassay containers per PCA replicate (15 containers total) at each location was stocked with 3600 g of fish

Table 9.4 Carrying capacities for lake trout in three water supplies determined by two different methods

| Hatchery site | $ECOC_{50}$ mg oxygen/L | Carrying capacity (kg/L/min) | |
		PCA	NH_3 accumulation
Wellsboro	10.2	6.2	13.7
Lamar	10.9	7.6	8.7
Leetown	5.8	3.0	3.5

Note: The two methods gave significantly different results at Wellsboro, but not at the other two sites (P < 0.05). The $ECOC_{50}$ is effective cumulated oxygen consumption below which fish fail to achieve at least 50% of their maximum growth.

and each 5-unit series was supplied with 3.0 L/min of water. Aeration was applied at the head of each rearing container so that influent DO concentrations were at least 90% of saturation. The bioassays were conducted for 4 to 8 weeks at each of the three sites.

At the end of each test, fish were weighed and the specific growth rates were calculated for each bioassay level. Specific growth was regressed against cumulative oxygen consumption and the $ECOC_{50}$ for each location was determined from the respective regression equations. Carrying capacities were calculated by computing the fish biomass, the oxygen consumption of which would equal the $ECOC_{50}$ value per L/min of water flow. For example, if a total fish load of 15 kg corresponded to an $ECOC_{50}$ of 9.0 mg/L DO in a flow of 3.0 L/min, the carrying capacity from the PCA procedure would be 5.0 kg per L/min.

The carrying capacity estimated from the $ECOC_{50}$ value was compared to that determined by estimated NH_3 accumulation for each of the three water supplies using Chi-square analysis at the 95% confidence level.

Calculated carrying capacities, in kg per L/min, based on estimated NH_3 accumulation to 0.016 mg/L were 13.7 at Wellsboro, 8.7 at Lamar, and 3.5 at Leetown (Table 9.4). Carrying capacities based on PCA bioassay determinations were 6.2 at Wellsboro, 7.5 at Lamar, and 3.0 at Leetown (Table 9.4). Calculated carrying capacities were nominally higher than those determined experimentally, but statistically different (P < 0.05) only for Wellsboro (Table 9.4). Calculated results consider only NH_3 toxicity, whereas the PCA method accounts for other unknown factors such as toxicity of other metabolites, the effects of chronic disease epizootics, and water quality factors influencing NH_3 toxicity. Thus, the expected PCA values would be lower than those calculated from estimated NH_3 exposure. The fact that the actual carrying capacity at Wellsboro was significantly different from that predicted from NH_3 calculations indicates that, for some water supplies, factors other than NH_3 exposure constitute important limitations to fish production.

SAMPLE PROBLEMS

1. A 1-Ha pond, 1 m deep, contains 4000 kg of tilapia being fed a 25% protein diet at 3% of their body weight per day. The concentration of TAN in the morning, before feeding, is 1.0 mg/L. What will be the TAN concentration 24 hours later?

2. The pH of a catfish pond changes during the day because of the photosynthetic removal of CO_2 and its subsequent replacement by respiration. Assume that the TAN level remains constant because the ammonia production rate equals the ammonia removal rate and the water temperature is a constant 28°C. At dawn the pH is 6.5 and the NH_3 concentration is 0.002 mg/L. At dusk the pH has risen to 9.5. To what level of NH_3 are the fish now being exposed?

3. How many kg of 30-cm rainbow trout can be reared in 500 L/min of 11°C water with a pH of 7.2? The maximum allowable NH_3 level is 0.016 mg/L, the protein content of the diet is 45%, and liquid oxygen is used.

4. The Department of Natural Resources of a Great Lakes state has declared that water supplies for production of 20 g Chinook salmon smolts must allow for at least three water uses with respect to DO in reaerated water. Calculate the maximum permissible pH of water for salmon hatcheries for an area where the water temperature is 9°C, the elevation is 185 m above sea level, and the maximum allowable NH_3 level is 0.0125 mg/L. Salmon are fed a 48% protein diet.

5. Consider a catfish farm in a western U.S. state using 2000 L/min of geothermal water. The pH is 9.5, the temperature is 30°C, and the elevation is 1500 m above sea level. Production raceways are built on a mountainside with gravity aeration between them. Assume that the vigorous agitation associated with the waterfall between raceways removes all the gaseous ammonia (NH_3) between raceways. The first set of raceways is loaded to capacity with 500 g fish being fed a 32% protein diet at 3% body weight per day. What is the NH_3 concentration at the head of the second set of raceways?

6. A national fish hatchery in the American West at an elevation of 600 m above sea level uses reservoir water with a pH of 6.8, heated to 14°C, to rear 20 g steelhead smolts on a 48% protein diet. The hatchery heats and discharges 20,000 L/min and circulates 200,000 L/min through the rearing ponds and aerators. Thus, the water is used 10 times before it is discharged. Calculate the NH_3 concentration at the hatchery effluent. Does this facility require the use of ammonia-removal filters to protect the fish from NH_3 toxicity?

7. A fish-rearing facility on the East Coast of the United States uses river water heated by an electric generating facility. The flow is 113 L/sec, the pH is 7.4, and the temperature is 13°C. Liquid oxygen

is injected along the raceway so that DO tensions are always above 150 mm Hg and rainbow trout are reared at a density of 176 kg/m³. The raceway is 27 m long, 2.5 m wide, and 1 m deep. Calculate the NH_3 concentration at the tail of the raceway.

8. Trout are grown in a serial water reuse system in order to evaluate a water source for a hatchery. Each of the rearing units is stocked with 4.5 kg of 6-cm fish and the series receives 8 L/min of water. After 5 weeks, influent and effluent DO concentrations of each rearing unit are measured and the fish are weighed. The following data are obtained:

Serial unit	Fish weight (kg)	Influent DO (mg/L)	Effluent DO (mg/L)
1	8.5	10	7.5
2	8.7	9.5	6.9
3	7.4	9.5	7.1
4	6.2	9.6	7.2
5	5.3	9.6	7.1

Calculate the $ECOC_{50}$ and the production potential of this water supply. When the hatchery is built, how many rearing units should be placed in a series?

9. Calculate the production potential of the water supply tested in Problem 8 using the ammonia accumulation procedure. The hatchery will produce 10-cm trout for stocking. The temperature is 11°C and the pH is 7.5.

10. A trout hatchery in Pennsylvania produces 25-cm rainbow trout for stocking recreational fisheries. The water temperature is 12°C and the fish diet contains 42% protein. Environmental regulations require that the hatchery discharge must not exceed 2.0 mg/L TAN. What is the maximum loading rate for this hatchery? How much could this production level be increased if the diet was replaced with one containing 38% protein with fat added so that the two diets were isocaloric?

References

Ball, I. R. 1967. The relative susceptibilities of some species of freshwater fish to poisons—I. Ammonia. *Water Research* 1: 767–775.

Bates, R. G. and G. D. Pinching. 1949. Acid dissociation constant of ammonia ion at 0 to 50°C and the base strength of ammonia. *Journal of Research of the National Bureau of Standards* 42: 419–430.

Brockway, D. R. 1950. Metabolite products and their effects. *Progressive Fish-Culturist* 12: 127–129.

Bullock, G. L. 1972. *Studies on Selected Myxobacteria Pathogenic for Fishes and on Bacterial Gill Disease of Hatchery-Reared Salmonids.* U.S. Bureau of Sport Fisheries and Wildlife, Technical Paper 60.

Burkhalter, D. E. and C. M. Kaya. 1977. Effects of prolonged exposure to ammonia on fertilized eggs and sac fry of rainbow trout, *Salmo gairdneri. Transactions of the American Fisheries Society* 106: 470–475.

Burrows, R. E. 1964. *Effects of Accumulated Excretory Products on Hatchery-Reared Salmonids.* U.S. Fish and Wildlife Service Research Report 66.

Colt, J. and G. Tchobanoglous. 1978. Chronic exposure of channel catfish, *Ictalurus punctatus*, to ammonia: Effects on growth and survival. *Aquaculture* 15: 353–372.

Emerson, K., R. C. Russo, R. E. Lund, and R. B. Thurston. 1975. Aqueous ammonia equilibrium calculations: Effects of pH and temperature. *Journal of the Fisheries Research Board of Canada* 32: 2379–2383.

Flis, J. 1968. Anatomicohistopathological changes induced in carp, *Cyprinus carpio*, by ammonia water, II. Effects of subtoxic concentrations. *Acta Hydrobiologica* 10: 225–233.

Haskell, D. C. 1959. Trout growth in hatcheries. *New York Fish and Game Journal* 6: 205–237.

Kawamoto, N. Y. 1961. The influences of excretory substances of fishes on their own growth. *Progressive Fish-Culturist* 35: 26–29.

Larmoyeaux, J. C. and R. G. Piper. 1973. Effects of water reuse on rainbow trout in hatcheries. *Progressive Fish-Culturist* 35: 2–8.

Lloyd, R. and L. D. Orr. 1969. The diuretic response by rainbow trout to sub-lethal concentrations of ammonia. *Water Research* 3: 35–344.

Lovell, T. 1989. *Nutrition and Feeding of Fish.* Van Nostrand, New York, New York.

Meade, J. W. 1985. Allowable ammonia for fish culture. *Progressive Fish-Culturist* 47: 135–145.

Meade, J. W. 1988. A bioassay for production capacity assessment. *Aquacultural Engineering* 7: 139–146.

Muir, J., J. van Rijn, and J. Hargreaves. 2000. Production in intensive and recycle systems. Pages 405–445 *in* M. C. M. Beveridge and B. J. McAndrews, editors. *Tilapias: Biology and Exploitation.* Kluwer Publishers, London, England.

Redner, B. D. and R. R. Stickney. 1979. Acclimation to ammonia by *Tilapia aurea. Transactions of the American Fisheries Society* 108: 383–388.

Robinette, H. R. 1976. Effects of selected sublethal levels of ammonia on the growth of channel catfish, *Ictalurus punctatus. Progressive Fish-Culturist* 38: 26–29.

Smart, G. 1976. The effects of ammonia exposure on the gill structure of rainbow trout, *Salmo gairdneri. Journal of Fish Biology* 8: 471–478.

Smith, C. E. and R. G. Piper. 1975. Lesions associated with the chronic exposure to ammonia. Pages 497–514 *in* W. E. Ribelin and G. Migali, editors. *The Pathology of Fishes.* University of Wisconsin Press, Madison, Wisconsin.

Snieszko, S. F. and G. L. Hoffman. 1963. Control of fish diseases. *Laboratory Animal Care* 13: 197–206.

Soderberg, R. W. 1985. Histopathology of rainbow trout, *Salmo gairdneri* (Richardson), exposed to diurnally fluctuating un-ionized ammonia levels in static-water ponds. *Journal of Fish Diseases* 8: 57–64.

Soderberg, R. W. 1992. Linear fish growth models for intensive aquaculture. *Progressive Fish-Culturist* 54: 255–258.

Soderberg, R. W. 1995. *Flowing Water Fish Culture.* Lewis Publishers, Boca Raton, Florida.

Soderberg, R. W., J. B. Flynn, and H. R. Schmittou. 1983. Effects of ammonia on the growth and survival of rainbow trout in intensive static-water culture. *Transactions of the American Fisheries Society* 112: 448–451.

Soderberg, R. W., M. V. McGee, and C. E. Boyd. 1984a. Histology of cultured channel catfish, *Ictalurus punctatus* Rafinesque. *Journal of Fish Biology* 24: 683–690.

Soderberg, R. W., M. V. McGee, J. M. Grizzle, and C. E. Boyd. 1984b. Comparative histology of rainbow trout and channel catfish in intensive static-water aquaculture. *Progressive Fish-Culturist* 46: 195–199.

Soderberg, R. W. and J. W. Meade. 1991. The effects of ionic strength on un-ionized ammonia concentration. *Progressive Fish-Culturist* 53: 118–120.

Soderberg, R. W. and J. W. Meade. 1992. Effects of sodium and calcium on acute toxicity of un-ionized ammonia to Atlantic salmon and lake trout. *Journal of Applied Aquaculture* 1: 83–92.

Thurston, R. V., R. C. Russo, and C. E. Smith. 1978. Acute toxicity of ammonia and nitrite to cutthroat trout fry. *Transactions of the American Fisheries Society* 197: 361–368.

Thurston, R. V., R. C. Russo, C. M. Fetterolf, Jr., T. A. Edsall, and Y. M. Barber, Jr., editors. 1979. *A Review of the EPA Red Book: Quality Criteria for Water*. Water Quality Section, American Fisheries Society, Bethesda, Maryland.

Walters, G. R. and J. A. Plumb. 1980. Environmental stress and bacterial infection in channel catfish, *Ictalurus punctatus* Rafinesque. *Journal of Fish Biology* 17: 177–185.

Willingham, W. T., J. E. Colt, J. A. Fava, B. T. Hillaby, C. L. Ho, M. Katz, R. C. Rosso, D. L. Swanson, and R. V. Thurston. 1979. Ammonia. Pages 6–18 *in* R. V. Thurston, R. C. Russo, C. M. Fetterolf, Jr., T. A. Edsall, and Y. M. Barber, Jr., editors. *A Review of the EPA Red Book: Quality Criteria for Water*. Water Quality Section, American Fisheries Society, Bethesda, Maryland.

chapter ten

Fish hatchery effluent control

Water pollution from aquaculture is currently under considerable scrutiny from governmental environmental authorities. Flowing water aquaculture effluents contain solid waste, dissolved solids, fertilizer nutrients, and BOD (biochemical oxygen demand) that may harm sensitive receiving waters, such as trout streams. Fish production at some facilities is effluent driven, meaning that carrying capacity is determined by the levels of pollutants discharged to the receiving waters. Effluent treatment to reduce total suspended solids (TSS), total nitrogen (TN), and total phosphorus (TP) appropriately to flow through trout aquaculture involves (1) improved farm management and (2) effluent treatment. Farm management strategies to improve effluent water quality include the provision of the best possible rearing conditions to avoid stress and nutrient excretion. Keeping dissolved oxygen (DO) levels above 100 mm Hg and maintaining high fish densities have been shown to reduce nutrient loads in trout hatchery effluents (Sindilariu 2007).

Solids removal in the quiescent zone

Solid waste is the pollutant of greatest concern in trout hatchery effluents and most recommendations for making hatcheries compliant with environmental regulations call for separation and removal of solid waste from the hatchery discharge.

Solids concentrations reported for trout hatchery discharges range from 5 to 50 mg/L and carry 7%–32% of TN, 30%–84% of TP (Cripps and Bergheim 2000), and about 85% of BOD (Willoughby et al. 1972). Thus, solids removal significantly improves the quality of aquaculture discharges.

Solids collection from aquaculture raceways in the United States most commonly relies on a zone at the rear from which fish are excluded by a screen (Hinshaw and Fornshell 2002). Solids control using a raceway quiescent zone (QZ) depends upon fish activity in the rearing zone keeping solids in suspension and then significant settling of this waste behind the screen of the relatively short, fishless QZ (Figure 10.1).

Solids are removed from the QZ by pumping them into a collection tank or flushing them to an off-line settling basin. This process is considered to be Best Management Practice (BMP) in the Idaho Waste Management Guidelines for Aquaculture Operation (IDEQ 1977). JRB

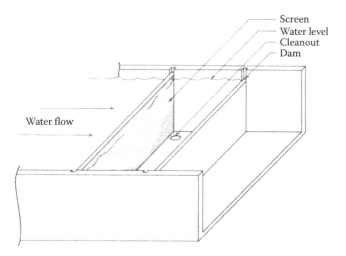

Figure 10.1 Detail of quiescent zone to capture solid waste at the end of a raceway (see Figure 4.3).

Associates (1984) determined that installation of QZs at the ends of all raceways of trout hatcheries was the best pollutant control technology. Factors affecting the efficiency of solids removal from the QZ probably include raceway water velocity, the size of the QZ, and fish density in the rearing area of the raceway above the QZ. Settling efficiency is theoretically independent of depth (Colt and Tomasso 2001). Raceway QZs are widely reported to capture between 75% (Colt and Tomasso 2001) and 90% (Hinshaw and Fornshell 2002) of suspended solids. These values trace back to Jensen (1972) and JRB Associates (1984), or subsequent authors who cited them.

Youngs and Timmons (1991) calculated that the minimum velocity for cleaning of fish feces was 2.4 cm/s, based on a specific gravity of 1.19 (Chen 1991). Minimum cleaning velocity for a mixture of fish feces and wasted feed was estimated to be 3.7 cm/s (Youngs and Timmons 1991). Since these are calculated values, they are pertinent to areas without fish.

Burrows and Chenoweth (1955) reported that the minimum raceway velocity to prevent settling of salmon solids within the rearing zone was 2.4–3.0 cm/s. Westers and Pratt (1977) and Boersen and Westers (1986) found that baffles were necessary to keep solids from settling in the raceways when the velocity was 3.3 cm/s. Hinshaw and Fornshell (2002) reported that, at velocities less than 3.96 cm/s, fecal solids and waste feed will accumulate wherever fish are not present.

Wong and Piedrahita (2003) described the QZ as consisting of the last 1–5 m of the raceway. MacMillan et al. (2003) recommended a 6.1 m QZ for very large raceways in the Snake River valley of Idaho. Boersen and

Westers (1986) recommended a 3 m² QZ for raceways 30 m long, 3 m wide, and 0.67 m deep.

The Idaho Department of Environmental Quality (DEQ) BMP recommends that QZs be cleaned twice per month from raceways with large fish loads and once per month from raceways with smaller fish loads. It has recently been found that nutrient leaching occurs rapidly from trout sludge. Stuart et al. (2006) reported that the majority of nutrients were released in the first 24 hours of deposition. True et al. (2004) found that a trout farm with a once per week QZ cleaning schedule had 49% of P in the dissolved form, compared to 67% dissolved P in the discharge of a hatchery with a QZ cleaning schedule of every 2 weeks. These reports make it clear that punctual removal of solids from settling basins and QZs could have a large impact on reducing P effluent loading. MacMillan et al. (2003) suggested that a cleaning schedule of once per week represented a reasonable compromise between labor requirements and effluent compliance.

Wong and Piedrahita (2000) evaluated solids settling efficiency in laboratory, fishless systems in terms of the overflow rate (OFR). The OFR is the volumetric flow rate divided by the surface area of the settling basin (QZ). For example, the OFR of a 22,308 cm² QZ at the end of a raceway receiving 25,233 cm³/s of water would be 1.13 cm/s. If the size of the QZ were doubled to 44,616 cm², the OFR becomes (25,233 cm³/s)/(44,616 cm²/s) = 0.56 cm/s. They estimated that a sedimentation basin with an OFR of 0.5 cm/s would capture about 81% of the settleable material. Idaho Waste Management Guidelines for Aquaculture (IDEQ 1997) recommend an OFR of 0.94 cm/sec for raceway QZs. The laboratory experiments of Wong and Piedrahita (2000) predicted a diminishing effect in solids removal efficiency as OFR is reduced. Predicted removal efficiencies ranged from 44% at an OFR of 4 cm/s to 91% at an OFR of 0.125 cm/s with predicted settling of 73% and 81% at OFR values of 1.0 cm/s and 0.5 cm/s, respectively (Wong and Piedrahita 2000).

The authors of IDEQ (1977) suggest that the required size of the QZ be determined by setting the OFR equal to the settling velocity of the solid waste, then doubling the value obtained. Solid waste from trout aquaculture is a complex mixture of whole and broken fecal casts, small fecal particles, and particles of various sizes of uneaten feed. Trout fecal casts have settling velocities of 2–5 cm/s, while fine, smaller particles have settling velocities of 0.046–0.09 cm/s (Warrer-Hansen 1982; Stechey and Trudell 1990). Settleable solids from a commercial trout farm had a median settling velocity of 1.7 cm/s (Wong and Piedrahita 2000). If settling velocity is assumed to be 1.7 cm/s and this is set equal to the desired OFR for a raceway receiving 25,233 cm³/s of water, the required size of the QZ is 25,233 cm³/sec/1.7 cm/s = 14,843 cm². Doubling this value to 29,686 cm² results in the QZ occupying the last 1.2 m of a 2.4 m wide raceway. The resulting OFR is 0.85 cm/s, which is close to the IDEQ (1977) recommendation of 0.98 cm/s.

Figure 10.2 Off-line settling lagoon to receive QZ cleaning waste. Sludge is removed from the basin and stored in the steel tank in the background until conditions for on-land application are suitable.

Cleaning waste from the QZs is sent to an off-line settling basin, often called a polishing lagoon (Figure 10.2), with a long enough retention time to reduce TSS to acceptable levels for discharge to receiving waters or to a second polishing lagoon that provides final solids settling for the entire hatchery water supply (Figure 10.3). Whole system polishing lagoons may

Figure 10.3 Polishing lagoon designed for final treatment of an entire hatchery water supply. Sludge collected is eventually applied on land.

not be appropriate for facilities that have thermal restrictions as part of their discharge criteria due to their size and resulting long water retention times. IDEQ (1997) recommends that polishing lagoons have an OFR of 0.046 cm/sec.

Fish activity in the rearing zone of an effective raceway keeps solids suspended and the absence of fish in the QZ allows for settling. Fish densities in trout hatcheries are generally equal in lb/ft^3 to one-half the fish length in inches (Piper et al. 1982), or 95 kg/m^3 for 30-cm fish, but Soderberg (1995) has shown that much higher densities are possible. The successful function of a raceway should depend upon fish density as well as water flow velocity and OFR of the QZ, but early studies of QZ efficiencies did not include fish density as a variable in this process.

Soderberg (2007) investigated the QZ sedimentation efficiency of trout raceways under different conditions of raceway velocity, OFR, and fish density. Efficiency was measured by comparing the weight of total suspended solids in the QZ overflow to that retained in the QZ.

Solids retention in the QZs was negligible at velocities of 2.0 and 4.4 cm/s regardless of fish density and at a fish density of 40 kg/m^3 regardless of velocity. Solids collection efficiencies were affected more by OFR and fish density than by velocity. Solids collection was significantly better at a fish density of 160 kg/m^3 than at a density of 80 kg/m^3 at OFR values of 0.5 and 1.0 cm/s. Solids retention was better at an OFR of 0.5 cm/s than at an OFR of 1.0 cm/s at a density of 80 kg/m^3, but was not affected by OFR at a fish density of 160 kg/m^3. Solids collection efficiency was not affected by velocity in the range of 2.4–3.4 cm/s at either density. Thus, optimum conditions for solids collection in trout raceway QZs were at a fish density of 160 kg/m^3 at OFR values of 0.5 or 1.0 cm/s, or at a fish density of 80 kg/m^3 and an OFR of 0.5 cm/s, and at raceway velocities from 2.4 to 3.4 cm/s. Under these conditions, up to 20% of the total solids load was collected in the raceway QZs, but the mean values did not exceed 13.1% (Table 10.1).

Table 10.1 Mean (SD) solids removal efficiency (% TSS trapped in QZ) at two fish densities, overflow rates (OFR) of 0.5 and 1.0 cm/s, at raceway velocities ranging from 2.4 to 3.4 cm/s in Wellsboro, Pennsylvania in 2005

Fish density (kg/m^3)	OFR (cm/s)	Percent TSS in QZ
80	0.5	10.4 (6.2) a
160	0.5	13.1 (7.2) a
80	1.0	7.1 (3.8) b
160	1.0	12.6 (5.8) a

Source: Data from Soderberg, R. W. 2007. *North American Journal of Aquaculture* 69: 275–280.

Note: Means followed by the same letter are not significantly different (P < 0.05).

Table 10.2 Total suspended solids (TSS) concentrations (SD) at effluents of raceways containing fish at two densities, with quiescent zones at two overflow rates (OFR) and at raceway velocities ranging from 2.4 to 3.4 cm/s in Wellsboro, Pennsylvania in 2005

Fish density (kg/m^3)	OFR (cm/s)	TSS (mg/L)
80	0.5	2.8 (1.7)
160	0.5	2.6 (1.3)
80	1.0	3.4 (2.8)
160	1.0	3.7 (2.7)

Source: Data from Soderberg, R. W. 2007. *North American Journal of Aquaculture* 69: 275–280.
Note: There were no differences among treatment means (P < 0.05).

Although this value is lower than expected from the literature, actual solids levels were less than 5 mg/L and solids concentrations in the effluents of downstream units of serial raceways were not additive (Table 10.2) (Soderberg 2007).

Raceway QZs are generally believed to be more efficient at trapping suspended solids than what was reported by Soderberg (2007). The value of 90% removal at a velocity of 3.3 cm/s reported by Jensen (1972) is widely cited (Westers and Pratt 1977; Hinshaw and Fornshell 2002). Jensen (1972), working in Michigan, reported secondhand information from the U.S. Fish and Wildlife Service National Fish Hatchery at Lamar, Pennsylvania and referred to settling of the wastes from four serial raceways in an entire 33 m raceway below them. This is not the same as settling individual raceway solids in a relatively small QZ. The Lamar, Pennsylvania study is reported in detail by McLaughlin (1981), who revealed that the 90% removal figure was determined by comparing influent and effluent concentrations of TSS. The sampling procedure is not described. There is no mention of continuous sampling and analysis of composite samples as was done in the study conducted by Soderberg (2007). Most of the other reports on the use of sedimentation to remove TSS from hatchery effluents involve large basins rather than short QZs (McLaughlin 1981; Mudrak 1981).

Much of the work resulting in QZ performance expectations is based on measurements of waste particle settling velocities in laboratory conditions (Wong and Piedrahita 2000). Chesness et al. (1975) found close agreement between predicted and actual solids removal in one experiment, but in another test, the predicted efficiency was 84.5% and the actual solids removal was only 57.3%.

JRB Associates (1984) measured the differences between influent and effluent TSS levels in raceway QZs at three commercial hatcheries in Idaho. This is the only study in which TSS levels were actually measured at the effluents of QZ-equipped raceways. At the Rangen Hatchery, raceway-effluent TSS concentrations averaged 7.32 mg/L in 1983, prior to installation of QZs. Quiescent zones were installed in 1984 and mean effluent TSS concentrations were 1.69 mg/L. They reported that installation of the QZs caused a 75% reduction in TSS discharged. This figure has been widely cited to justify BMP for trout aquaculture. At the Jones Hatchery, screened and unscreened raceways were compared in the same year. Effluents from raceways with and without QZs had mean TSS levels of 2.31 mg/L and 3.97 mg/L, respectively. This represents a 42% reduction in TSS attributed to the QZs. At Crystal Springs Hatchery, the TSS levels were 5.62 mg/L and 3.25 mg/L for unscreened and screened raceways, respectively, or a reduction of 42% TSS attributed to the QZs (JRB Associates 1984).

One reason for differences in results from Soderberg (2007) and that conducted by JRB Associates (1984) could be variation in diets. Hatchery managers are well aware that different diets produce waste particles of varying size and density (IDEQ 1977; Mudrak 1981; Hinshaw and Fornshell 2002). The diets in use in 1984 were much different from the high-density diet used by Soderberg (2007).

Soderberg (2007) showed that raceway QZs trap less solid waste from raceway effluents than was previously reported, perhaps indicating the need for further experiments evaluating effects of diet types and the addition of coagulants to trout diets (Brinker et al. 2005).

The results presented by Soderberg (2007) introduce the question of what level of solids retention is necessary for compliance with pollution discharge regulations and what measures are necessary to achieve these limits. In serial raceways, with fish at high densities and OFR values of 1.0 cm/s, mean TSS levels were only 2.6 mg/L even though more than 80% of the solids load was being lost to the receiving water (Soderberg 2007). Hatcheries that are unable to comply with present TSS limitations probably have QZs that trap much less solid waste than is reported for optimum conditions described by Soderberg (2007), meaning that their discharges could be brought into compliance by manipulating fish density, water velocity, and OFR.

If the discharge limitation for TSS is much less than 5.0 mg/L, QZ technology may not be sufficient to maintain trout hatchery compliance regardless of hydraulic and biological conditions. Early studies of sedimentation of trout hatchery solids indicate that solids retention can be considerably improved by expanding the sedimentation basin to the size of an entire raceway. It is doubtful that this would be a suitable procedure due to difficulty in cleaning such a large basin without resuspending and losing a large fraction of the collected solid waste.

Many references support the recommendation of solids removal to maximize P removal from effluents, but others report significant leaching of P from sludge stored in the clarifiers. Removal of dissolved phosphorus is difficult but, when required, could be accomplished by granular or mechanical filtration, biological treatment, or chemical precipitation. These advanced treatment options may be practical for treating microscreen backwash water, but for large flows there are few economically viable phosphorus removal options beyond the adoption of good solids capture and disposal practices (Summerfelt and Vinci 2008).

Production and control of phosphorus

Phosphorus is an environmental pollutant associated with fish hatchery effluents because of its effect on the eutrophication of hatchery receiving waters (see Chapter 18). The obvious solution to reducing phosphorus in hatchery effluents is to reduce the amount of phosphorus in fish diets. Modern, low-phosphorus diets contain a minimum of 1% phosphorus. Green et al. (2002) reported that 31% of dietary phosphorus is retained in the fish, the remainder being released as waste. Thus, the phosphorus production from a fish hatchery can easily be estimated from the fish feeding rate.

Estimates of the total phosphorus associated with the solids fraction of fish wastes range from 30% to 84% (Cripps and Bergheim 2000) to 40% (True et al. 2004). Thus, efficient solids capture can substantially reduce hatchery effluent phosphorus levels. Summerfelt and Vinci (2008) stated that because phosphorus is primarily associated with solids, rapid solids removal from the system is the best way to remove phosphorus from recirculating aquaculture effluents. Removal of dissolved phosphorus from hatchery effluents is not practical, but chemical coagulation is a possibility for future efforts to control hatchery effluent phosphorus levels.

Rishel and Eberling (2006) investigated the use of alum and various synthetic polymers to treat the backwash water from an intensive aquaculture system. Total suspended solids, dissolved P, total P, and TN were reduced by 99%, 92%–99%, 98%, and 87%, respectively. Actual concentrations, after chemical treatment, were 4–20 mg/L TSS, 0.16 mg/L dissolved P, and 0.9–3.0 mg/L TP. The tests were conducted in jars so no recommendation on how the chemicals would be added to and mixed in a clarifier was given.

Primary sedimentation of municipal wastewaters removes only 10%–15% of P. If chemicals are added, effluent concentrations can be reduced to 0.8–1.5 mg P/L. If the effluent is filtered, the concentration can be reduced to 0.2 mg/L (Balmer and Hultman 1988).

Use of coagulants and flocculants for bulk-flow treatment of aquaculture effluents has not been documented in the scientific literature, but these reports indicate that chemical treatment of aquaculture effluents should be further investigated for situations with extreme P discharge limits. Sindilariu (2007) estimated that the chemical cost to treat 1000 gpm (3785 L/min) would be $59/day.

Production and control of nitrogen

Ammonia is the principal nitrogenous product of trout metabolism and is toxic to fish in its un-ionized form (see Chapter 9). Ammonia toxicity is therefore of importance to life in hatchery receiving waters and is usually regulated as such. Nitrogen is also often considered a source of eutrophication in receiving waters even though there is little evidence confirming this supposition (see Chapter 18). Nonetheless, total nitrogen discharged is often regulated as well as ammonia. Ammonia discharge limitations should be based on NH_3 levels, but are often based on total ammonia nitrogen (TAN) concentrations. Total ammonia discharged into low pH receiving waters has a much smaller environmental effect than the same amount of TAN discharged into receiving waters with a higher pH (see Chapter 9). Green et al. (2002) reported that 85% of the total nitrogen in hatchery discharges is in the ammonia species. Thus, the concentrations of total nitrogen and TAN in hatchery discharges are easily estimated. Cripps and Bergheim (2000) found that 7%–32% of total nitrogen in hatchery discharges was in the solids fraction. As a result, effective treatment of TSS reduces effluent levels of nitrogen.

Mechanical filtration

Mechanical filters such as disc filters (Figures 10.4 and 10.5) and drum filters (Figure 11.7), also known as microscreen filters, can remove a large portion of the solid waste from fish hatchery effluents. These units are very expensive but may be required to meet strict environmental standards imposed by pollution control agencies.

Microscreen filters are available from several manufacturers and operate by passing water through a fine-mesh screen. In the case of a drum filter, the screen is in the form of a cylinder. A disc filter retains solids with a series of circular, vertical screens. Disc filters are preferred over drum filters for treating large water volumes, such as total system flows, due to their greater hydraulic capacities. Drum filters are usually selected for treatment of QZ or dual-drain tank waste, where smaller water volumes are involved. Water passes from the inside to the outside of the screen. When pores clog, as determined by a water level difference across the screen, the backwash mechanism deploys and solids are washed from the screen into a waste discharge line, that leads to an off-line settling basin for final treatment (Figure 10.2). During typical operation, the screen is washed from several hundred to several thousand times per day (Libey 1993; Summerfelt et al. 1999). Compared to other methods of suspended solids removal, microscreen filters have high hydraulic capacities, low space requirements, and small water loss to waste disposal. Summerfelt (1999) reported that microscreen backwash flows range from 0.2% to 1.5%

Figure 10.4 Microscreen disc filter treating an entire hatchery water flow in conventional fish hatchery application. Each unit receives 2800 L/min.

Figure 10.5 Cover removed from disc filter to show detail of the screens.

of total system flows. Schematic diagrams of drum and disc filters are found in Timmons and Ebeling (2013).

Typical screen openings used in aquaculture range from 40 to 100 μm. In this screen size range, TSS removal can be from 30% to 80% (Timmons and Ebeling 2013). Brinker et al. (2005) reported a filtration efficiency of 83% using an 80 μ screen. Screen sizes smaller than 30 μ lead to excessive backwash frequency and do not increase filtration efficiency (Timmons and Ebeling 2013). Cripps (1995) and Kelly et al. (1997) found that screen sizes below 60–100 μ did not improve filtration efficiency over larger pore sizes. The efficiency of microscreens is significantly better at high solids loadings than when solids loads are lighter (Summerfelt 1999). Thus, they are well suited to handling concentrated effluents such as the bottom discharges from dual drain tanks or QZ cleaning waste from raceways.

SAMPLE PROBLEMS

1. Explain how the often cited, but incorrect, value of 90% TSS removal for raceway QZs resulted from a failure of the scientific process.
2. For a raceway that is $15 \, m \times 2.5 \, m \times 0.6 \, m$, receiving 1000 L/min, where should the screen be placed so that OFR = 0.5 cm/s? Where should it be placed so that OFR = 1.0 cm/s?
3. For the above raceway, if the screen groove is 1.2 m from the dam and cannot be adjusted (QZ dimensions of $2.5 \, m \times 0.6 \, m$), what water flow should be selected to achieve an OFR of 0.5 cm/s? 1.0 cm/s?
4. Estimate the concentration of total phosphorus (TP) in the effluent from a hatchery containing 100,000 kg of trout being fed 1.0% BW/day if the hatchery raceways contain efficient quiescent zones (QZs).
5. Estimate the TP concentration in the effluent of the hatchery above if it is equipped with microscreen filters in addition to the QZs.
6. Estimate the total nitrogen (TN) concentrations in the effluent described in Problem 3.
7. Estimate TN concentration in the effluent described in Problem 4.
8. Hatchery production limits are sometimes effluent driven. What is the production limit for a hatchery with efficient QZs whose discharge limit for TP is 1200 kg/year?
9. What is the fish production limit for the hatchery in Problem 8 if all of the discharge passes through microscreen filters?
10. Design an off-line settling basin to treat 500 L/min of QZ cleaning waste.

References

Balmer, P. and B. Hultman. 1988. Control of phosphorus discharges: Present situation and trends. *Hydrobiologia* 170: 305–319.

Boersen G. and H. Westers. 1986. Waste solids control in hatchery raceways. *Progressive Fish-Culturist* 48: 151–154.

Brinker, A., W. Koppe, and R. Rosch. 2005. Optimizing trout farm effluent treatment by stabilizing trout feces: A field trial. *North American Journal of Aquaculture* 67: 244–258.

Burrows, R. and H. Chenoweth. 1955. *Evaluation of Three Types of Fish Rearing Ponds*. Research Report 39, U.S. Department of the Interior, Fish and Wildlife Service, Washington, DC.

Chen, S. 1991. Theoretical and experimental investigation of foam separation applied to aquaculture. PhD Thesis. Cornell University, Ithaca, New York.

Chesness, J. L., W. H. Poole, and T. K. Hill. 1975. Settling basin design for raceway fish production systems. *Transactions of the American Society of Agricultural Engineers* 18: 159–162.

Colt, J. E. and J. R. Tomasso. 2001. Hatchery water supply and treatment. Pages 91–186 *in* G. A. Wedemeyer, editor. *Fish Hatchery Management*. Second edition, American Fisheries Society, Bethesda, Maryland.

Cripps, S. J. 1995. Serial particle size fractionation and characterization of an aquacultural effluent. *Aquaculture* 133: 323–339.

Cripps, S. J. and A. Bergheim. 2000. Solids management and removal for intensive land-based aquaculture production systems. *Aquacultural Engineering* 22: 33–56.

Green, J. A., R. W. Hardy, and E. L. Brannon. 2002. Effects of dietary phosphorus and lipid levels on utilization and excretion of phosphorus and nitrogen by rainbow trout (*Oncorhynchus mykiss*). 1. Laboratory-scale study. *Aquaculture Nutrition* 8: 279–290.

Hinshaw, J. and G. Fornshell. 2002. Effluents from raceways. Pages 77–104 (Chapter 4) *in* J. R. Tomasso, editor. *Aquaculture and the Environment in the United States*. U.S. Aquaculture Society, a Chapter of the World Aquaculture Society, Baton Rouge, Louisiana.

IDEQ (Idaho Division of Environmental Quality). 1977. *Idaho Waste Management Guidelines for Aquaculture Operations*. Boise, Idaho.

Jensen, R. 1972. Taking care of wastes from the trout farm. *American Fishes and U.S. Trout News* 16: 4–6, 21.

JRB Associates. 1984. *Development of Effluent Limitations for Idaho Fish Hatcheries*. Submitted to the U.S. Environmental Protection Agency, Office of Water Enforcement (EPA Contract 68-01-6514), Washington, DC.

Kelly, L. A., A. Bergheim, and J. Stellwagon. 1997. Particle size distribution of wastes from freshwater fish farms. *Aquaculture International* 5: 65–87.

Libey, G. S. 1993. Evaluation of a drum filter for removal of solids from a recirculating aquaculture system. Pages 529–532 *in* J-K. Wang, editor. *Techniques for Modern Aquaculture*. American Society of Agricultural Engineers, St. Joseph, Michigan.

MacMillan, J. R., T. Huddleston, M. Woolley, and K. Fothergill. 2003. Best management practice development to minimize environmental impact from large flow-through trout farms. *Aquaculture* 226: 91–99.

McLaughlin, T. W. 1981. Hatchery effluent treatment, U.S. Fish and Wildlife Service. Pages 167–173 *in* L. J. Allen and E. C. Kinney, editors. *Proceedings of the Bio-Engineering Symposium for Fish Culture*. American Fisheries Society, Bethesda, Maryland.

Mudrak, V. A. 1981. Guidelines for economical commercial fish hatchery wastewater treatment systems. Pages 174–182 *in* L. J. Allen and E. C. Kinney, editors. *Proceedings of the Bio-Engineering Symposium for Fish Culture*. American Fisheries Society, Bethesda, Maryland.

Piper, R. G., I. B. McElwain, L. E. Orme, J. P. McCraren, L. G. Fowler, and J. R. Leonard. 1982. *Fish Hatchery Management*. U.S. Department of the Interior, Fish and Wildlife Service, Washington, DC.

Rishel, K. L. and J. M. Eberling. 2006. Screening and evaluation of alum and polymer combinations as coagulation/flocculation aids to treat effluents from intensive aquaculture systems. *Journal of the World Aquaculture Society* 37: 191–199.

Sindilariu, P. 2007. Reduction in effluent nutrient loads from flow-through facilities for trout production: A review. *Aquaculture Research* 38: 1005–1036.

Soderberg, R. W. 1995. *Flowing Water Fish Culture*. Lewis Publishers, Boca Raton, Florida.

Soderberg, R. W. 2007. Efficiency of trout raceway quiescent zones in controlling suspended solids. *North American Journal of Aquaculture* 69: 275–280.

Stechey, D. and Y. Trudell. 1990. *Aquaculture Wastewater Treatment: Wastewater Characterization and Development of Appropriate Treatment Technologies for the Ontario Trout Production Industry: Final Report*. Canadian Aquaculture Systems Bioengineering Technologies & Business Management Services, Queens Printer for Ontario, Canada.

Stuart, N. T., G. D. Boardman, and L. A. Helfrich. 2006. Characterization of nutrient leaching rates from settled rainbow trout (*Oncorhynchus mykiss*) sludge. *Aquacultural Engineering* 35: 191–198.

Summerfelt, S. T. 1999. Waste-handling systems. Pages 309–350 *in* F. Wheaton, editor. *CIGR Handbook of Agricultural Engineering*, Vol. II. Aquacultural Engineering, American Society of Agricultural Engineers, St. Joseph, Michigan.

Summerfelt, S. T., P. R. Adler, D. M. Glenn, and R. Kretschmann. 1999. Aquaculture sludge removal and stabilization within created wetlands. *Aquacultural Engineering* 19: 81–92.

Summerfelt, S. T. and B. J. Vinci. 2008. Better management practices for recirculating aquaculture systems. Pages 389–426 *in* C. S. Tucker and J. A. Hargreaves, editors. *Environmental Best Management Practices for Aquaculture*. Wiley-Blackwell, Ames, Iowa.

Timmons, M. B. and J. M. Ebeling. 2013. *Recirculating Aquaculture, Third Edition*. Northeastern Regional Aquaculture Center Publication No. 401-3013. Ithaca, New York.

True, B., W. Johnson, and S. Chen. 2004. Reducing phosphorus discharge from flow-through aquaculture I: Facility and effluent characterization. *Aquacultural Engineering* 32: 129–144.

Warrer-Hansen, I. 1982. Methods of treatment of waste water from trout farming. Pages 113–121 *in* J. S. Alabaster, editor. *Report of the EIFAC Workshop on Fish Farm Effluents*. EIFAC (European Inland Fisheries Advisory Commission) Technical Paper No. 41. FAO, Rome, Italy.

Westers, H. and K. M. Pratt. 1977. Rational design of hatcheries for intensive salmonid culture, based on metabolic characteristics. *Progressive Fish-Culturist* 39: 157–165.

Willoughby, H., H. N. Larsen, and J. T. Bowen. 1972. The pollutional effects of fish hatcheries. *American Fishes and U.S. Trout News* 1: 6–20.

Wong, K. B. and R. H. Piedrahita. 2000. Settling velocity characterization of aquacultural solids. *Aquacultural Engineering* 21: 233–246.

Wong, K. B. and R. H. Piedrahita. 2003. Solids removal from aquacultural raceways. *World Aquaculture* 34(1): 60–67.

Youngs, W. D. and M. B. Timmons. 1991. A historical perspective of raceway design. Pages 160–169 *in* Engineering Aspects of Intensive Aquaculture. Northeast Regional Agricultural Engineering Service Publication NRAES-49. Cooperative Extension, Ithaca, New York.

chapter eleven

Water recirculation

The most intensive level of aquaculture is fish production in recirculating systems that recycle all, or nearly all, of the water through the fish rearing units. Recirculating systems are one of two general types depending upon the amount of water that is recycled. The so-called water reuse systems were developed by U.S. governmental agencies to accelerate the production of salmonid smolts by using heated water. Water was reused at a rate of 90% per flow cycle in order to conserve energy used in water heating. The flow through the system is therefore 90% used water and 10% new, heated water. This is equivalent to using the water 10 times in a serial raceway system with reaeration between each of the raceways. Water reuse hatcheries use biological filters to remove ammonia, but this level of treatment may not be necessary in low-pH waters (Figure 11.1).

Recirculating Aquaculture Systems (RAS) refer to those facilities where virtually all the water is recycled and the water requirement is only that necessary to replace losses due to evaporation, spillage, and filter backwash. RASs are used within buildings to maintain suitable water temperatures for the species of fish selected for culture. The technology of fish culture in an RAS involves meeting the following environmental requirements of the fish in a self-contained unit:

1. Maintenance of water temperatures for optimum fish growth
2. Maintenance of dissolved oxygen (DO) tensions suitable for efficient respiration and fish health
3. Maintenance of safe levels of NH_3 and other toxic metabolites
4. Maintenance of water reasonably clear of solid waste

Maintenance of suitable DO tensions is addressed in Chapter 8. The present chapter describes the filtering of ammonia and solid waste from the recirculating stream. Ammonia can be removed from water by pH elevation followed by air stripping, by ion exchange resins or zeolites, or by biological filtration. Aquaculture applications rely on biological filtration in which ammonia is biologically oxidized to nitrate. The total nitrogen level does not change, but toxic ammonia is converted to nitrate, which is nontoxic to fish.

Figure 11.1 Extent to which water may be reused without ammonia removal following aeration is dependent upon the pH and its effect on ammonia ionization.

Biological filtration

Microbiology

Biological filtration depends upon nitrification, a microbial process by which autotrophic bacteria oxidize ammonium (NH_4^+) to nitrite (NO_2^-) and then to nitrate (NO_3^-). Ammonium is relatively nontoxic to fish, but its removal from culture water reduces levels of its equilibrium product, ammonia (NH_3). The intermediate product, NO_2^-, is quite toxic to fish and is an important concern in recirculating aquaculture. Nitrate is essentially nontoxic to fish and is allowed to accumulate in recirculating systems.

Several genera of bacteria are known to oxidize ammonium and nitrite, but the most important are *Nitrosomonas* for oxidizing NH_4^+ to NO_2^- and *Nitrobacter* for the subsequent oxidation to NO_3^-. The nitrification rate is influenced by pH, DO, bicarbonate and ammonia levels, and temperature.

The literature on pH optima for *Nitrosomonas* and *Nitrobacter* are conflicting (Hochheimer 1990), probably because these bacteria adapt to the pH conditions to which they are exposed. Biological filters can probably operate from as low as pH 5 to as high as pH 10 if bacteria are allowed to acclimate to these conditions and pH changes are not too abrupt (Wheaton et al. 1991).

The stoichiometric requirement for the oxidation of 1 g of ammonia-N is 4.34 g of oxygen, and DO levels of at least 2 mg/L are recommended to assure that nitrification is not oxygen limited (Sharma and

Albert 1977; Manthe et al. 1985; Wheaton et al. 1991). Nitrification reduces pH by consuming alkalinity and producing hydrogen ions. One gram of ammonia conversion to nitrate requires 7.15 g of alkalinity ($CaCO_3$), and alkalinity values of 75–150 mg/L $CaCO_3$ (Gujer and Boller 1986; Allain 1988) must be maintained to ensure optimal operation of a biological filter. Recirculating aquaculture systems require the addition of a buffer, usually sodium bicarbonate, to maintain suitable pH and alkalinity levels (Hochheimer 1990). Base is generally metered in the system flow as necessary to maintain desired water chemistry. Bisogni and Timmons (1991) present a diagram for estimating the amount of base required to maintain alkalinity, and thus pH, based on losses due to nitrification. According to their estimates, approximately 0.11 kg of $NaHCO_3$ would be required per kg of food fed per day.

Un-ionized ammonia is toxic to nitrification bacteria and may inhibit the nitrification process. *Nitrobacter* is inhibited by NH_3 levels as low as 0.1–1.0 mg/L, while *Nitrosomonas* is more tolerant (Anthonisen et al. 1976). Nitrous acid (HNO_2), which is in equilibrium with NO_2^- in aqueous solution, is toxic to nitrification bacteria at concentrations as low as 0.22 mg/L (Anthonisen et al. 1976), but the equilibrium constant between HNO_2 and NO_2^- is so high ($10^{-3.14}$) that virtually no HNO_2 is present at pH values greater than 5.0. Thus, nitrite toxicity is of little concern in recirculating aquaculture as a nitrification inhibitor. It is, however, an important water quality factor affecting fish health in these systems.

Nitrification bacteria are found in nature in a wide variety of environmental conditions and thus can adapt to a wide range of temperatures. Nitrification is most efficient at temperatures from 30°C–35°C, but nitrification occurs from –5°C–42°C (Jones and Morita 1985; Laudelout and Van Tichelen 1960) and nitrification bacteria will adapt to a wide range of temperatures. Most workers report that the nitrification rate is proportional to temperature (Knowles et al. 1965; Haug and McCarty 1971; Speece 1973; Soderberg 1995), but Zhu and Chen (2002) showed no influence of temperature on nitrification rates at temperatures from 14°C–27°C. Malone and Pfeiffer (2006) reported that temperature is no longer considered to be an important factor in controlling nitrification rates in biofilters, but Wortman and Wheaton (1991) showed a small but significant temperature effect and suggested the following equation to adjust nitrification rates in the temperature range of 7°C–35°C:

$$R = 140 + 8.5\,T\,(°C) \tag{11.1}$$

where R is the relative nitrification rate. This equation predicts that nitrification at 10°C would be 72% that at 20°C. Discrepancies in the reported effects of temperature on nitrification are probably primarily due to the wide biodiversity of nitrification bacteria, which grants their ability to colonize diverse environments.

Speece (1973) and Soderberg (1995) provide simple expressions for nitrification rates which result in predictions on the order of 1.5 g N/m²/day, but the data used for these predictive equations were obtained in experimental conditions without fish. Fish waste contains organic material that results in colonies of heterotrophic bacteria that compete with the nitrification autotrophs for filter surface space. The organic load on a biofilter is affected by the efficiency of solids removal, and thus is variable in RASs. Losordo (1997) reported that between 0.25 and 0.55 g/m²/day is a more reasonable range for expected nitrification rates in production-scale biofilters. Malone et al. (1993) reported nitrification rates of 0.280 g/m²/day for a rotating biological contactor (RBC) and 0.284 g/m²/day for a fluidized bed in actual aquaculture environments. The actual nitrification rate of a particular biofilter depends upon the efficiency of organic matter control and the allowable ammonia level, which is a function of system pH. Timmons and Ebeling (2013) reported nitrification rates of 0.2–1.0 g/m²/day at temperatures of 15°C–20°C and 1.0–2.0 g/m²/day at temperatures from 25°C–30°C, demonstrating the variability in biofilter performance due to application-specific factors. Malone et al. (1993) and Losordo (1997) both worked with warmwater fish, so for warmwater aquaculture applications, 0.28 g/m²/day appears to be a reasonable starting point for biofilter design. This value can be adjusted for temperature using Equation 11.1. For example, if 0.28 g/m²/day is accepted as a design criterion for 28°C water, the adjusted value for 10°C water would be 0.17 g/m²/day.

Filter start-up is the most critical period in RASs because ammonia reaches high levels while the *Nitrosomonas* colony is being established. Similarly, NO_2^- reaches high levels following the ammonia peak until the *Nitrobacter* population expands to accommodate its oxidation. Because of this, biofilters are usually started with inorganic ammonia at up to about 15 mg/L (Hochheimer 1990), and microbial populations are allowed to become established before fish are stocked. Alternatively, a filter may be started with media from an active biofilter.

Biofilter configuration

Biofilters are designed by matching the ammonia production of the fish to the nitrification capacity of the filter. Biofilter media are characterized by their specific surface areas, which is the area of surface available for microbial colonization per unit of media volume. Primitive biofilters originally used for fish culture contained stone or plastic rings with specific surface areas of 60–130 m²/m³ (Soderberg 1995) (Table 11.1) (Figure 11.2). When it was found that low specific surface area media did not provide

Table 11.1 Specific surface areas of some biofilter media types
used in aquaculture

Media type	Specific surface (m²/m³)
2.5–7.5-cm stone	63
9-cm plastic rings	89–102
4-cm plastic rings	131
Rigid plastic module	89
8-mm polyethylene pellets	656
5-mm polyethylene beads	1150
3-mm polyethylene beads	1475
10-mm polystyrene beads	403
3-mm polystyrene beads	1260
1-mm polystyrene beads	3936
Graded sand	4000–45,000

sufficient nitrification, some biofilters were modified by replacing them
with media that had a greater specific surface area (Figure 11.3).

The biological filter is a housing containing media with a large surface area that becomes colonized by the nitrification bacteria. Water from
the recirculation stream passes through the filter containing the media
and the attached microbial flora. A wide variety of media types and filter
configurations has been investigated since the early 1970s, but the present state-of-the-art filter utilizes small beads to maximize the surface

Figure 11.2 Early aquaculture biofilters with low specific surface area media
such as these 9-cm plastic rings did not provide sufficient surface for complete
nitrification.

Figure 11.3 This biofilter originally held oyster shell media (63 m²/m³), which was replaced with 8-mm polyethelene pellets (656 m²/m³) to improve its nitrification ability. Water enters the filter housing from the bottom, floods up through the media, which is neutrally buoyant, and exits through ports above the media bed.

available for microbial colonization. The rotating biological contactor is also a commonly selected biofilter type for aquaculture.

In fluidized bed filters, water flows up or down through the media with a sufficient pressure to expand, or fluidize it. The media particles are thus held in motion so that they cannot make continuous contact with each other and the entire particle surface is available for bacterial colonization (Losordo et al. 1999). Floating bead biofilters (Figure 11.4), the most commonly used type of expandable media biofilter, use 3–5 mm polyethylene beads with specific surface areas of 1150–1475 m²/m³ (Malone et al. 1993). Floating bead biofilters operate by forcing water through the fine media, which accomplishes some solids filtration while simultaneously providing a large surface area for nitrification (Timmons and Ebeling 2013). Trapped solids and excessive bacterial growth are removed by periodically backwashing the filter.

Fluidized sand biofilters have also been investigated for aquaculture (Summerfelt 1999; Summerfelt et al. 2001). Specific surface area estimates for sand range from 4000–45,000 m²/m³ and thus promise to be much smaller than bead media filters. Fluidized sand biofilters have high energy requirements to fluidize the sand and are difficult to maintain, due to clogging by solids and trouble maintaining water flows and pressures that effectively fluidize the sand bed (Timmons and Ebeling 2013).

Microbead filters (Figure 11.5) use 1–3 mm (1260–3936 m²/m³) polystyrene beads that are highly buoyant compared to the neutral-density

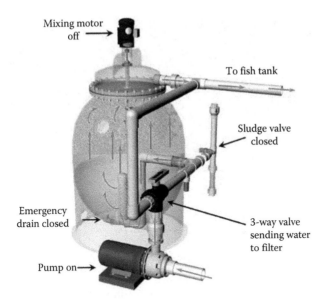

Figure 11.4 Floating bead biofilter.

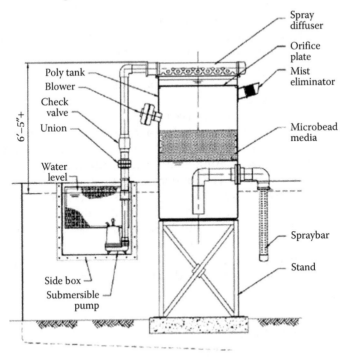

Figure 11.5 Microbead biofilter. (From Timmons, M. B. et al. 2006. *Aquacultural Engineering* 34: 332–343. With permission.)

Figure 11.6 Rotating biological contactor.

polyethylene beads used in floating bead filters. Influent water is distributed over the top of the media bed, trickles down through the media, and is discharged from the filter by gravity (Timmons and Ebeling 2013).

New generation biofilters containing small bead media provide much more surface for bacterial colonization than those originally designed for aquaculture. An additional advantage is that they occupy much less floor space, which is at a premium in indoor facilities.

The RBC filter is a series of disks on a shaft that rotates slowly in a trough of water (Figure 11.6).

Approximately 40% of the disk media is submerged (Losordo et al. 1999) at any one time, but its rotation alternates the disks between the ammonia-rich water and the air. The recycle-stream passes through the trough, parallel to the rotating shaft. The nitrification surface area of an RBC is the total surface area of both sides of all the disks in the unit. Losordo et al. (1999) described an RBC that used blocks of plastic media having a specific surface area of 200 m^2/m^3 rather than flat or corrugated disks. This media increased the nitrification capacities of traditional RBC designs, but the authors warned that the filter medium increases in weight as much as 10-fold due to the growth of the microbial flora, and the support structure of the unit must be designed to accommodate this weight (Soderberg 1995; Losordo et al. 1999).

Removal of solid waste

In water reuse hatcheries, sedimentation provided by raceway quiescent zones (QZs) is adequate to keep water sufficiently free of turbidity, but the accumulation of recirculating solid waste in RASs must be controlled by

Figure 11.7 Microscreen drum filter in an RAS application.

mechanical filters. Solids removal has been the most important technical obstacle to the success of RASs, but the recent availability and widespread use of microscreen filters has made control of solid waste in RASs less of a problem (Figure 11.7).

Microscreen filters have a separate solids waste stream that must be managed to result in a complete waste management system. The waste stream is the backwash. Backwash flow ranges from 0.2%–1.5% of the treated flow (Summerfelt et al. 1999), depending on screen opening size, frequency of backwash, and influent solids load to the filter. The backwash flow is generally directed to an off-line settling pond (Brazil and Summerfelt 2006) from which it can eventually be returned to the RAS. As in serial reuse systems, screen size should be greater than 60 m because smaller screen sizes lead to excessive backwash frequency and subsequent water loss in RASs (Timmons and Ebeling 2013).

RAS and water reuse system management

Any aquaculture system that employs biological filters would require a water chemistry laboratory capable of performing daily or continuous determinations of TAN, NH_3, NO_2^-, pH, total alkalinity, DO, and possibly CO_2 and chloride levels.

The components described for in-tank settling of solids, oxygenation, biofiltration, maintenance of pH and alkalinity, and suspended solids screening form the backbone of a state-of-the-art RAS or reuse system. Other components that may or may not be necessary for successful fish production in these systems include CO_2 stripping, water chemistry manipulation to control nitrite toxicity, ozonation, and foam fractionation.

Control of CO_2

Carbon dioxide can accumulate in aquaculture systems to levels that interfere with respiration. Furthermore, CO_2 decreases pH, increasing the need for water chemistry manipulation. Losordo (1997) recommends that CO_2 levels be kept below 25 mg/L to ensure fish health. Since the solubility of CO_2 at fish growth temperatures ranges from 0.4–1.0 mg/L, it would seem obvious that the use of liquid oxygen for aeration would prevent CO_2 levels from approaching 25 mg/L, but this is not necessarily true.

Predicting the quantity of CO_2 removed by aeration is difficult because composition of the gas phase varies due to its reaction with bicarbonate in water. The problem with quantifying CO_2 is complicated by the difficulty in accurately measuring hyper-supersaturated levels. The titration process removes some of the gas before it is measured, and calculation of CO_2 from alkalinity and pH is not reliable because the bicarbonate system would not be in equilibrium at times of extreme CO_2 levels.

Excess CO_2 can be removed from water by contacting it with air or oxygen or by the addition of NaOH to convert CO_2 to bicarbonate (Vinci et al. 1996), but the level of removal that is required, if any, is site-specific and should be evaluated during system operation.

Control of nitrite toxicity

Biofilter effluents contain residual levels of NO_2^-, and nitrite exposure causes a disease in fish called methemoglobinemia, in which functional anemia results from the conversion of hemoglobin to methemoglobin. Sublethal levels of NO_2^- increase the susceptibility of fish to bacterial diseases (Hanson and Grizzle 1985). One of the dangers in using biofilters in reuse aquaculture when they are not needed to control ammonia is that they produce NO_2^-, possibly causing more toxicity problems than if the existing ammonia were left untreated. Nitrite toxicity is related to environmental chloride levels, apparently because the gill epithelium cannot distinguish between Cl^- and NO_2^- and Cl^- inhibits NO_2^- absorption. Coho salmon exposed to 9 mg/L NO_2^- with a chloride concentration of 20 mg/L averaged a 64% conversion of their hemoglobin to methemoglobin, which resulted in 50% mortality. When fish were exposed to the same concentration of NO_2^- in a solution containing 148 mg/L Cl^-, none of the fish died and the incidence of methemoglobin conversion was 39%. Fish in this study survived an NO_2^- exposure of 30 mg/L when the chloride concentration was 261 mg/L (Perrone and Meade 1977). Losordo (1997) recommended that NO_2^--N levels be kept below 5 mg/L, but channel catfish exposed to this level for 24 hours showed a conversion of 77% of their hemoglobin to methemoglobin (Tomasso et al. 1979). Nitrite exposures as low as 1.0 mg/L resulted in a 21% conversion of hemoglobin to methemoglobin in channel catfish (Tomasso et al. 1979). Schwedler et al. (1985) measured the amount of methemoglobin in channel catfish exposed to NO_2^- at NO_2^-:Cl^- ratios of 1:1 and 1:3. Fish in the low Cl^- concentration had 80% methemoglobin, but there was only 25% methemoglobin in the fish at the higher Cl^- concentration.

Nitrite toxicity is also related to calcium concentrations, probably because divalent cations decrease membrane permeability (Potts and Fleming 1974), thereby reducing NO_2^- absorption. Steelhead trout in solutions containing 67 mg/L Cl^- from NaCl exhibited a 96 hr LC_{50} (median lethal concentration) of 0.97 mg/L NO_2^-. When the same amount of Cl^- was added as $CaCl_2$, the 96-hr LC_{50} of NO_2^- to steelhead trout was 22.7 mg/L (Wedemeyer and Yasutake 1978).

It is impossible to assign a maximum tolerable concentration for NO_2^- to fish because its toxicity is highly variable depending upon pH, chloride, and calcium levels. Nitrite toxicity can be controlled by increasing the volume of biofilter media, increasing ozone to oxidize NO_2^- to NO_2^-, or adding chloride. Recommendations on effective NO_2^-:Cl^- ratios range from 1:3 (Schwedler et al. 1985) to 1:20 (Losordo 1997). Because Ca^{2+} protects fish from NO_2^- toxicity, $CaCl_2$ is more effective than NaCl for this purpose. Because of the effect of chloride on NO_2^- toxicity, fish in brackish or saline waters are resistant to methemoglobinemia (Almendras 1987).

Foam fractionation and ozone

In-tank settling removes solid waste particles of >100 um in size, and microscreen filters used for aquaculture most often have pore sizes of 40 um. Smaller particles, including dissolved organic matter and colloids, may become important in aquaculture where all, or nearly all, of the water is recirculated. These fine particles may be harmful to fish and provide a substrate for heterotrophic bacteria, which compete for space with the nitrification fauna of the biofilter, thereby reducing its performance. The requirement for mechanisms to remove this component of the solids load is controversial, but when elected, ozone and/or foam fractionation are the processes of choice.

Ozone is a powerful oxidizing agent that, when added to water, reduces NO_2^- to NO_3^- and oxidizes dissolved organic matter and colloidal solids. Ozone-treated water enhances fine solids removal by foam fractionation (Losordo 1997), improving the performance and reducing backwash frequency of microscreen filters (Summerfelt et al. 1996). Because of its unstable nature, ozone cannot be stored and must be generated on site. Ozone is most commonly added along with oxygen during oxygenation, but care must be taken to not expose fish to toxic levels. Wedemeyer et al. (1979) reported gill damage to fish at ozone levels as low as 5 ug/L. Summerfelt et al. (1996) recommended that ozone be added to the oxygen feed line at 3%–4% of the gas flow in the amount of 25 g/kg food. Brazil et al. (1996), however, reported that 13 g/kg of food was a sufficient ozone dose for fine solids control. Bullock et al. (1996) found that 25 g ozone/kg food was safe for fish when added in the oxygen feed line, but 36–39 g/kg occasionally caused toxicity to fish.

Foam fractionation works by the tendency of fine organic particles to adhere to gas bubbles, forming skimmable suds, or foam. Such a foam is sometimes produced in the oxygenation process, but foam fractionation is usually accomplished by devices specifically designed for the process that vigorously mix air into the water to be treated in a closed chamber. These units have a convenient mechanism for collection and removal of the foam.

Summary

Successful RASs are presently applicable to facilities that produce larval or juvenile fish because of the low fish loads and high value by weight of the fish produced (Figures 11.8 and 11.9).

RASs can hypothetically produce food fish (tilapia) at an estimated cost of $3.57/kg (Lutz 2000) and have the potential for profitable operation by supplying ethnic live fish markets. However, these markets are rather limited and easily saturated, and they are not presently competitive with Asia or Central America for processed tilapia (Costa-Pierce 2000). The potential future of food fish RAS aquaculture in temperate, developed locations lies in marketing campaigns to portray fresh, locally produced fish as a high-quality product worthy of premium prices.

At least 13 large RASs in the United States, intended to produce food fish, have failed (Timmons and Ebeling 2013). The most commonly reported reason for the lack of success of these ventures was system failure, but inadequate biofilters, state environmental regulations, poor management, lack of operating capital, high production costs, inadequate fish growth rate, disease, and inadequate water supply were also listed as causes of failure (Timmons and Ebeling 2013).

SAMPLE PROBLEMS

Solve Problems 1 through 7 concerning the RAS culture of 4000 kg of 600 g tilapia in water heated to 30°C. The flow of recirculated water through the system is 1000 L/min.

1. What surface area of biofilter media is required? What volume of media is required if a floating bead biofilter is selected?
2. Design an RBC filter for this application using 6 cm thick plastic sheeting to construct the disks. How would this design change if blocks with a specific surface area 200 m²/m³ were substituted for the plastic disks?

Figure 11.8 Two views of an RAS used to produce a wide variety of larval and juvenile sport fish for stocking recreational fisheries in New Jersey. Heated water is mixed with ambient temperature water to provide optimum temperatures for each species of fish produced.

Figure 11.9 Two views of an RAS in Xining, China, used to produce juvenile whitefish (*Coregonus* sp.) for stocking in grow-out cages in the Yellow River.

3. What volume of media is required if a fluidized bed filter or micro-bead filter is selected?

4. Water entering the biofilter contains 100 mm Hg of DO. Calculate the DO concentration of the biofilter effluent if no oxygen is added during passage through the filter. Is aeration of the biofilter required?

5. How much sodium bicarbonate ($NaHCO_3$) must be continuously metered into the system to replace alkalinity destroyed by nitrification?

6. The residual nitrite level is 5 mg/L, and fish are exhibiting clinical signs of methemoglobinemia. Calculate the concentration of $CaCl_2$ that must be maintained in the system to control NO_2^- toxicity.

7. Oxygen requirements for the system will be met with bulk liquid oxygen. How much oxygen must be delivered per month, assuming an absorption efficiency of 90%?

8. Consider 90% reuse systems for the production of 20 cm steelhead smolts. Water is heated to 15°C in order to complete the growth of these fish in 1 year. The hatcheries will be built in the western United States where the elevation is 500 m above sea level. Calculate the pH above which biological filters will be required.

9. A hatchery in the midwestern United States circulates 10,000 L/min of reused water, heated to 12°C to produce 20,000 kg of 20-cm lake trout. The water supply is a lake with a pH of 7.2 at an elevation of 200 m above sea level. If no biofilters are used, how much new water must be continuously heated and added to the recirculating flow?

References

Allain, P. A. 1988. Ion shifts and pH management in high density shedding systems for blue crabs (*Callinectus sapidus*) and red swamp crayfish (*Procambarus clarkii*). Master's Thesis, Louisiana State University, Baton Rouge, Louisiana.

Almendras, J. M. E. 1987. Acute nitrite toxicity and methemoglobinemia in juvenile milkfish, *Chanos chanos* Forsskal. *Aquaculture* 61: 33–40.

Anthonisen, A. C., R. C. Loehr, T. B. S. Prakasam, and E. G. Srinath. 1976. Inhibition of nitrification by ammonia and nitrous acid. *Journal Water Pollution Control Federation* 48: 835–852.

Bisogni, J. J., Jr. and M. B. Timmons. 1991. Control of pH in closed cycle aquaculture systems. Pages 333–348 *in Engineering Aspects of Intensive Aquaculture*. Northeast Regional Agricultural Engineering Service, Cornell University, Ithaca, New York.

Brazil, B. L. and S. T. Summerfelt. 2006. Aerobic treatment of gravity thickening tank supernatant. *Aquacultural Engineering* 34: 92–102.

Brazil, B. L., S. T. Summerfelt, and G. S. Libey. 1996. Application of ozone to recirculating aquaculture systems. Pages 373–389 *in* G. S. Libey and M. B. Timmons, editors. *Successes and Failures in Commercial Recirculating Aquaculture*. Northeast Regional Agricultural Engineering Service, Ithaca, New York.

Bullock, G. L., S. T. Summerfelt, A. C. Noble, A. L. Weber, M. D. Durant, and J. T. Hankins. 1996. Effects of ozone on outbreaks of bacterial gill disease and numbers of heterotrophic bacteria in a trout culture recycle system. Pages 598–610 *in* G. S. Libey and M. B. Timmons, editors. *Successes and Failures in Commercial Recirculating Aquaculture*. Northeast Regional Agricultural Engineering Service, Ithaca, New York.

Costa-Pierce, B. A. 2000. Challenges facing the expansion of tilapia aquaculture. Preface *in* B. A. Costa-Pierce and J. E. Rakocy, editors. *Tilapia Aquaculture in the Americas*, Vol. 2. The World Aquaculture Society, Baton Rouge, Louisiana.

Gujer, W. and M. Boller. 1986. Design of a nitrifying tertiary trickling filter based on theoretical concepts. *Water Research* 20: 1353–1362.

Hanson, L. A. and J. M. Grizzle. 1985. Nitrite-induced predisposition of channel catfish to bacterial diseases. *Progressive Fish-Culturist* 47: 98–101.

Haug, R. T. and P. L. McCarty. 1971. *Nitrification with Submerged Filter*. Technical Report Number 149. Department of Civil Engineering, Stanford University, Stanford, California.

Hochheimer, J. N. 1990. Trickling filter model for closed system aquaculture. Unpublished Dissertation, University of Maryland, College Park, Maryland.

Jones, R. D. and R. Y. Morita. 1985. Low temperature growth and whole cell kinetics of a marine ammonium oxidizer. *Marine Ecology Progress Series* 21: 239–243.

Knowles, G., A. L. Downing, and M. J. Barrett. 1965. Determination of kinetic constants for nitrifying bacteria in mixed culture with the aid of an electronic computer. *Journal of General Microbiology* 79: 263–278.

Landelout, H. and L. Van Tichelin. 1960. Kinetics of nitrite oxidation by *Nitrobacter winogradsky*. *Journal of Bacteriology* 79: 39–42.

Losordo, T. M. 1997. Tilapia culture in intensive recirculating systems. Pages 185–211 *in* B. A. Costa-Pierce and J. E. Rakocy, editors. *Tilapia Aquaculture in the Americas*, Vol. 1. The World Aquaculture Society, Baton Rouge, Louisiana.

Losordo, T. M., M. B. Masser, and J. E. Rakocy. 1999. *Recirculating Aquaculture Tank Production Systems: A Review of Component Options*. Publication No. 453. Southern Regional Aquaculture Center, College Station, Texas.

Lutz, C. G. 2000. Production economics and potential competitive dynamics of commercial tilapia culture in the Americas. Pages 119–132 *in* B. A. Costa-Pierce and J. E. Rakocy, editors. *Tilapia Aquaculture in the Americas*, Vol. 2. The World Aquaculture Society, Baton Rouge, Louisiana.

Malone, R. F., B. S. Chitta, and D. G. Drennan. 1993. Optimizing nitrification in bead filters for warmwater recirculating aquaculture systems. Pages 315–325 *in* J-K. Wang, editor. *Techniques for Modern Aquaculture*. American Society of Agricultural Engineers, St. Joseph, Michigan.

Malone, R. F. and T. J. Pfeiffer. 2006. Rating fixed film nitrifying biolfilters used in recirculating aquaculture systems. *Aquacultural Engineering* 34: 389–402.

Manthe, D. P., R. F. Malone, and H. Perry. 1985. Water quality fluctuations in response to variable loading in a commercial blue crab shedding system. *Journal of Shellfish Research* 3: 175–182.

Perrone, S. J. and T. L. Meade. 1977. Protective effect of chloride on nitrite toxicity to coho salmon, *Oncorhynchus kisutch*. *Journal of the Fisheries Research Board of Canada* 34: 486–492.

Potts, W. T. W. and W. R. Fleming. 1974. The effects of prolactin and divalent ions on the permeability to water of *Fundulus kansae*. *Journal of Experimental Biology* 53: 317–327.

Schwedler, T. E., C. S. Tucker, and M. H. Beleau. 1985. Non-infectious diseases. Pages 497–541 *in* C. S. Tucker, editor. *Channel Catfish Culture.* Elsevier, New York, New York.

Sharma, B. and R. C. Albert. 1977. Nitrification and nitrogen removal. *Water Research* 11: 897–925.

Soderberg, R. W. 1995. *Flowing Water Fish Culture.* Lewis Publishers, Boca Raton, Florida.

Speece, R. E. 1973. Trout metabolism characteristics and the rational design of nitrification facilities for water reuse in hatcheries. *Transactions of the American Fisheries Society* 102: 323–334.

Summerfelt, S. T. 1999. Waste-handling systems. Pages 309–350 *in* E. H. Bartali and F. Wheaton, editors, CIGR Series. *CIGR Handbook of Agricultural Engineering,* Volume II. American Society of Agricultural Engineers, St. Joseph, Michigan.

Summerfelt, S. T., P. R. Adler, D. M. Glenn, and R. Kretschmann. 1999. Aquaculture sludge removal and stabilization within created wetlands. *Aquacultural Engineering* 19: 81–92.

Summerfelt, S. T., J. Bebak-Williams, and J. Tsukuda. 2001. Controlled systems: Water reuse and recirculation. Pages 285–395 *in* G. Wedemeyer, editor. *Fish Hatchery Management.* Second edition. American Fisheries Society, Bethesda, Maryland.

Summerfelt, S. T., J. A. Hankins, A. L. Weber, and M. D. Durant. 1996. Effects of ozone on microscreen filtration and water quality in a recirculating rainbow trout culture system. Pages 163–172 *in* G. S. Libey and M. B. Timmons, editors. *Successes and Failures in Commercial Recirculating Aquaculture.* Northeast Regional Agricultural Engineering Service, Ithaca, New York.

Timmons, M. B. and J. M. Ebeling. 2013. *Recirculating Aquaculture.* Third edition. Publication No. 401-2013. Northeastern Regional Aquaculture Center, Ithaca, New York.

Timmons, M. B., J. L. Holder, and J. Ebeling. 2006. Applications of microbead biological filters. *Aquacultural Engineering* 34: 332–343.

Tomasso, J. R., B. A. Simco, and K. B. Davis. 1979. Chloride inhibition of nitrite induced methemoglobinemia in channel catfish *Ictalurus punctatus. Journal of the Fisheries Research Board of Canada* 36: 1141–1144.

Vinci, B. J., S. T. Summerfelt, M. B. Timmons, and B. J. Watten. 1996. Carbon dioxide control in intensive aquaculture: Design tool development. Pages 399–418 *in* G. S. Libey and M. B. Timmons, editors. *Successes and Failures in Commercial Recirculating Aquaculture.* Northeast Regional Agricultural Engineering Service, Ithaca, New York.

Wedemeyer, G. A., and W. T. Yasutake. 1978. Prevention and treatment of nitrite toxicity in juvenile steelhead trout, *Salmo gairdneri. Journal of the Fisheries Research Board of Canada* 35: 822–827.

Wedemeyer, G. A., N. C. Nelson, and W. T. Yasutke. 1979. Physiological and biochemical aspects of ozone toxicity to rainbow trout (*Salmo gairdneri*). *Journal of the Fisheries Research Board of Canada* 36: 605–614.

Wheaton, F. W., J. N. Hochheimer, G. E. Kaiser, and M. J. Krones. 1991. Principles of biological filtration. Pages 1–31 *in Engineering Aspects of Intensive Aquaculture.* Northeast Regional Agricultural Engineering Service, Cornell University, Ithaca, New York.

Wortman, B. and F. W. Wheaton. 1991. Temperature effects on biodrum nitrification. *Aquacultural Engineering* 10: 183–285.

Zhu, S. and S. Chen. 2002. The impact of temperature on nitrification rate in fixed film biofilters. *Aquacultural Engineering* 26: 221–237.

section two

Static water fish culture

chapter twelve

Static water fish culture

When fish are reared in a static pool of water, control of environmental factors affecting their growth and health is much reduced over the use of flowing water technology. Dissolved oxygen (DO), rather than being a steady component of the water supply, changes radically following a 24-hour cycle of photosynthesis and respiration. Fish are the only consumers of dissolved oxygen in flowing water systems, but in static water fish are a minor component of a complex oxygen budget. Water temperatures usually vary with the season since they are controlled by ambient air temperatures. Thus, static water aquacultures have a particular growing season based on a chosen fish species' temperature tolerances and seasonal surface water temperatures.

In flowing water fish culture, waste removal is accomplished by the flushing action of the water supply. In a static water pool, on the other hand, waste materials accumulate and must be metabolized by natural biological and chemical processes.

Carrying capacity, or the weight of fish that can be accommodated in a flowing water aquaculture facility, is determined by the volume of water flow and its proportional ability to supply oxygen and flush waste from the system. Carrying capacity in static water aquaculture is determined by dynamics of limnological processes in which nutrient levels affect photosynthesis and respiration rates of the resultant plant biomass which, in turn, affect dissolved oxygen fluctuations and the ability of the system to support fish.

It would seem intuitive to abandon static water aquaculture technologies for the apparently more modern and efficient flowing water methods, and many have made this mistake. Careful study reveals that flowing water technologies are not appropriate for most aquaculture applications. The first obvious obstacle is the requirement for complete, high quality diets for flowing water fish culture. In a static water pond, some fish food is naturally present. Consequently, some fish production will occur without management inputs. In flowing water fish culture, all food must be externally supplied and in many cases the cost of the artificial diet will be too high in relation to the value of the intended fish crop for the activity to be economically viable.

Early in my career, on assignment with the U.S. Peace Corps, I studied aquaculture technology for an indigenous *clarias* catfish that had a high market value. The fish was obviously carnivorous so my coworkers and I set out to manufacture a high-protein artificial diet. The least expensive fish in the market sold for two units of the local currency per kg. One kg of fresh fish weighed 300 g when dried. Thus, our cost for fish meal was 6.67 monetary units per kg. We mixed this meal with equal amounts of available carbohydrates (rice bran) and produced a primitive fish diet with an ingredient value of 4.50/kg. The catfish devoured our concoction with gusto and converted it to fish flesh at a rate of 2 kg of food per kg of gain. Thus, the net production cost for our fish was 9.00 units/kg. Unfortunately, the retail value of the fish was 8.00/kg. Later investigations showed that this fish could be produced economically at high densities in static water ponds enriched with manure that formed the ecological base for natural food organisms for the fish.

Flowing water fish culture requires expensive, complete diets. The positive economics of these aquacultures is due to the huge monetary differences between the cost of fish meal and the value of fish in the developed world. Such a differential does not exist in the developing world and the use of artificial diets are, thus, usually inappropriate for domestic markets.

In the past decade, the availability and use of artificial dry diets has led to dramatic increases in the aquaculture production of finfish and marine shrimp. This development has led to the elevation of farm-raised fish to global commodities.

Another obstacle to the widespread adoption of flowing water aquaculture involves water temperature and its effect on fish growth. Flowing water fish culture developed in the temperate zone for the production of salmonid fishes whose growth temperatures serendipitously coincide with groundwater temperatures in their native range. Since water supplies for salmonid hatcheries are usually springs or wells and groundwater is isothermal, or nearly so, fish grow year around. Water supplies outside the trout-growing regions of the world are not likely to be an appropriate temperature for growth of the desired fish species. For example, flowing water methods used to produce trout in the northern United States have not been adopted for catfish production in the south. The reason is that groundwater, even in the southern United States, is too cold for satisfactory growth of warmwater fishes. Rearing the fish in static water ponds allows for solar warming of the water in summer, which provides a growing season of sufficient duration for economic production.

Static water techniques provide the vast majority of aquaculture production worldwide. Global aquaculture production is approximately 158 million tonnes, nearly half of the total world fish production. At least 95% of this total is produced using static water technologies.

chapter thirteen

Review of limnology

Limnology, the study of the biological, physical, and chemical characteristics of lakes, forms the scientific background for static water aquaculture. Fish culture in stagnant pools can, in fact, be accurately described as applied limnology because it is the practice of manipulating limnological functions for a desired outcome in fish production. The purpose of the present chapter is to present a review of limnological principles pertinent to static water aquaculture.

Density of water and lake stratification

Water is at its maximum density at 4°C (32°F), solidifies at 0°C (32°F), and vaporizes at 100°C (212°F). These values account for some unique properties of water that determine its physical state on Earth, its thermal characteristics, and its value as a life-support medium.

Most of the Earth's water (97.3%) is in the oceans, and 81% of what is left is in the solid phase of ice. A trace is in the gas phase of vapor at any given time. This leaves about 7,039,000 cubic km of liquid, nonoceanic water on Earth, 6,935,000 km^3 of which is fresh water. Of the world's fresh water, 98% is groundwater, leaving a meager 127,300 km^3 in the liquid state on the Earth's surface.

The relationship of density to temperature is shown in Figure 13.1. Note that cold, liquid water is denser than ice and that the density of water decreases at an accelerating rate at temperatures above 4°C. The relationship shown in Figure 13.1 means that warm water floats on cold water, ice floats on cold water, and cold water is much denser than warm water. The most obvious observation to be gleaned from this is that lakes do not freeze solid in the temperate zone and thus support animal life. Limnology would be a much less interesting subject than it is if water reached its maximum density upon arriving at the solid phase, as do other compounds.

The fact that the density–temperature relationship for water is not linear causes lakes in the temperate zone to thermally stratify in the summer. When lake ice melts in the spring, the slush at the surface is near 0°C, and the bottom of the lake is 4°C because that is where the heaviest water has settled. Rapid solar heating causes the lake's surface to warm, and a heated layer of water, approximately equal in depth to the limit of light

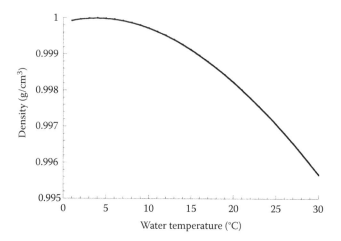

Figure 13.1 Relationship of density and temperature of pure water at 1 atmo-sphere of pressure. (Data from Cole, G. A. 1994. *Textbook of Limnology.* Waveland Press, Long Grove, Illinois.)

penetration, forms. The heated layer floats upon the cold water below, and there is a sharp temperature transition between the two layers.

Because warm water is so much lighter than cold water, the two layers do not mix until autumn, when atmospheric cooling reduces the tempera-ture differential. The warm top layer of a lake is called the epilimnion, the cold bottom layer is called the hypolimnion, and the rather shallow transition zone is called the thermocline. The epilimnion corresponds closely to the photic zone, which is the volume of the lake where plant production occurs. The hypolimnion receives insufficient light to conduct photosynthesis.

Trophic status of lakes

Lakes are broadly classified as being one of two types, eutrophic or oli-gotrophic. The distinction between the two types is one of fertility, with eutrophic lakes being nutrient rich and oligotrophic lakes being nutrient poor. The greater fertility of eutrophic lakes leads to higher levels of pri-mary productivity than in oligotrophic lakes.

Some of the most important principles of limnology and aquaculture relate to the abundance of dissolved oxygen in water and waterlogged sediments. The most notable distinction between eutrophic and oligo-trophic lakes is that the hypolimnia of eutrophic lakes becomes anoxic (devoid of dissolved oxygen) during the summer due to respiration and decomposition of aquatic organisms. Dissolved oxygen is retained in the hypolimnia of oligotrophic lakes because the low fertility results in a productivity level insufficient to deplete the oxygen supply before it is replenished by the fall turnover.

Waterlogged sediments, the mud bottoms of lakes and ponds, are always anoxic, but when the water in contact with the mud contains dissolved oxygen, as in oligotrophic lakes, there is a thin aerobic layer of mud at the water–sediment interface.

Anoxic water and sediments contain reduced molecular species. Nitrogen occurs principally as ammonium (NH_4^+) and nitrogen gas (N_2), sulfur occurs as sulfide (H_2S and FeS), and carbon compounds are reduced to methane (CH_4). Aerobic conditions favor oxidized species, principally nitrate (NO_3^-), sulfate (H_2SO_4, $CaSO_4$, $MgSO_4$), and oxidized carbon species (CO_2, HCO_3^-, and CO_3^{2-}). Of particular importance in the oxidation and reduction states of compounds in aquatic systems is the chemistry of iron and its effect on the solubility of phosphorus. In aquatic conditions, iron occurs in the ferric (Fe^{2+}) state, and in aerobic conditions, iron is in the ferrous (Fe^{3+}) state. Ferric iron compounds are black, and ferrous compounds are reddish, like rust. The oxidation state of pond muds can readily be observed by careful excavation. The top surface in contact with dissolved oxygen will be reddish or brownish in color. Less than 1 cm below this layer, the mud will be black, indicative of the presence of ferric iron compounds. The odiferous bubbles that arise from the disturbed anaerobic layer are made up of the gases hydrogen sulfide (H_2S), methane, (CH_4) and nitrogen (N_2).

Phosphorus is the nutrient most likely to limit primary production in aquatic systems. In fact, one of the characteristics defining the trophic status of lakes is the concentration of phosphorus. Ferric-phosphorus compounds are much more soluble than ferrous-phosphates. Thus phosphorus is much more abundant in reduced conditions (hypolimnia and surface sediments of eutrophic lakes) than in oxidized conditions (hypolimnia and surface sediments of oligotrophic lakes). This explains the common occurrence of sudden algal blooms associated with the fall turnover in oligotrophic lakes, prolific growth of aquatic macrophytes whose roots penetrate anaerobic sediments, and the richness of rivers below reservoirs with hypolimnetic discharges.

Another important result of the presence or absence of dissolved oxygen is the way that organic decomposition is affected. Anaerobic decomposition is much less efficient than decomposition in the presence of oxygen. Aerobic decomposition allows for the possibility of complete oxidation of organic matter to CO_2 and H_2O. In the absence of oxygen, anaerobic organisms only partially decompose organic matter, leaving some solid material behind. Incomplete oxidation of organic matter in waterlogged sediments is the origin of fossil fuels and the cause of natural lake aging. Lakes naturally mature from oligotrophic to eutrophic to swamps to wetlands and eventually to dry land. This is because they fill up with the products of aerobic decomposition. The carbon contained in these ancient sediments is in the form of gas, oil, coal, and peat and is trapped there until the fossil fuel is burned.

Dissolved oxygen dynamics in relation to trophic status

One of the first activities of pioneer limnologists was to measure dissolved oxygen profiles in lakes. In fact, the shape of the profile determines whether a lake is eutrophic or oligotrophic (Figure 13.2). Fish ponds are analogous to hypereutrophic lakes because the management applied to increase fish production (fertilization and/or feeding) increases nutrient levels beyond what limnologists normally observe in natural waters (Figure 13.2).

As eutrophication increases due to aquaculture management inputs, the depth of the photic zone decreases due to self-shading by the resultant plankton bloom. This decreases the volume of the epilimnion and increases the volume of the hypolimnion, thus reducing the space available to fish and increasing the volume of dangerously anoxic water below them.

Energy transfer in lakes

Photosynthesis is the basis of all life on Earth, and the photosynthetic fixation of carbon is called primary production. Organisms that conduct

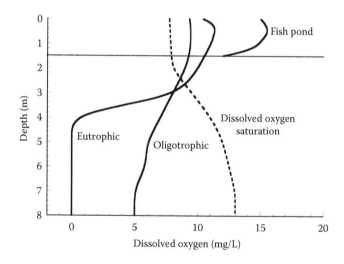

Figure 13.2 Afternoon dissolved oxygen profiles in typical eutrophic and oligotrophic lakes and in a typical static water fish pond. Note that surface waters are supersaturated and deep waters are undersaturated with dissolved oxygen.

primary production, primarily phytoplankton in aquatic systems, are called autotrophs. Heterotrophs are organisms that consume the products of primary production. Energy that enters ecological systems that is produced by autotrophs outside the system is referred to as allochthonous.

Inputs of allochthonous energy are especially important to fish production in natural waters and fish ponds. For example, shaded headwater trout streams receive little solar radiation and thus support an insignificant level of primary production. The food webs that lead to trout production are nearly completely dependent upon the allochthonous input of the very leaves that limit photosynthesis. The fallen leaves that enter the stream are photosynthate produced outside the stream. The bacteria that colonize the dead leaves may be considered first-order heterotrophs. An assemblage of insects and other invertebrates consume the bacteria-enriched leaf material. Carnivorous insects consume the herbivores, and all macroinvertebrates nourish the trout.

The biology of lakes

The biological components of lakes and ponds consist of the littoral (edge), plankton (suspended), benthic (bottom), detritus (dead and decomposing material and its associated fauna), and nekton (motile invertebrates and fish). The study of these components and their interactions is the science of aquatic ecology. Manipulation of these communities to increase production of a particular fish species, or group of species, is the science of static water aquaculture.

In natural lakes, emergent littoral plants such as cattails, sedges, and rushes and submergent littoral plants such as pond weeds (*Potamogetonaceae*) and macrophytic algae (chiefly *Chara* and *Nitella*) form an important component of the ecosystem. The functions of littoral macrophytes are much reduced in fish ponds, and these plants are generally considered a nuisance because few organisms feed on them, thus stalling transfer of their considerable energy stores to useful trophic levels. Similarly, the energy in coarse filamentous algae such as *Spirogyra* and *Pithophora* is not efficiently transferred to useful fish biomass in aquaculture, and management is usually applied to avoid these plants.

The phytoplankton, on the other hand, are primary producers that form the base of the food web that leads most efficiently to the production of many important aquaculture fish species. Management efforts to avoid the proliferation of macrophytes and filamentous algae divert plant nutrients, CO_2, and sunlight to the more desirable phytoplankton.

Also included in the phytoplankton are photosynthetic (*Cyanobacteria*) and nonphotosynthetic bacteria. The *Cyanobacteria*, or blue-green algae, are primary producers, but the nonphotosynthetic forms are heterotrophs that convert dissolved organic matter into particulate forms that are efficiently transferred to higher trophic levels by grazers such as ciliates, flagellates, and zooplankton.

The zooplankton community is composed primarily of protozoans, rotifers, cladocerans, and copepods. Some protozoans are photosynthetic but most are consumers of bacteria and minute phytoplankton, thus forming an important link to higher trophic levels.

The microcrustacean rotifers, cladocerans, and copepods are the primary food items for many larval and a few adult fish species important in aquaculture. Some microcrustaceans function as primary consumers by filtering plankton from the water. Others are carnivorous, feeding on other zooplankton or protozoans.

The littoral benthos of lakes is a complex and varied community of plants and animals including algae, protozoans, rotifers, cladocerans, copepods, gastropods, nematodes, leeches, and insects. Littoral animals are in close proximity to the plankton on which they feed. In deeper regions of lakes, the benthic community is much reduced over that observed in the littoral zone. This is especially true in eutrophic lakes where only a few species of insects, worms, and clams can tolerate the low oxygen conditions of the hypolimnion. Fish ponds are generally designed so as not to contain a sublittoral zone (Figure 13.2).

Detritus is the heterotrophic community that decomposes dead material and organic matter in various stages of decomposition. Dead material is much more abundant in aquatic systems than living biomass. Organic matter is broken down into dissolved and particulate forms chiefly by the action of bacteria and fungi. Bacteria are consumed by members of the zoobenthos, particularly ciliated protozoans which in turn are consumed by predatory zooplankton and insects. The detritus community is a very important link in the transfer of energy in aquatic systems. Some important aquaculture fish species feed directly on detritus.

Fish species selected for aquaculture include members of all piscine ecological niches. Silver carp (*Hypophthalmichthys molitrix*) and most tilapia (*Oreochromis*) species are filter feeders of phytoplankton. Bighead carp (*Aristichthys nobilis*), paddlefish (*Polyodon spathula*), and many larval fish species selectively graze on zooplankton. Some tilapias, notably *Tilapia melanopleura, T. rendalli,* and *T. zilli,* and grass carp (*Ctenopharyngodon idella*) are herbivores of macrophytes. Mullets (*Mugil* spp.), milkfish (*Chanos chanos*), and common carp are detritivores while ictalurids and salmonids selectively target macroinvertebrates. Basses (*Micropterous* spp. and *Morone* spp.) are piscivores in their adult forms.

Applied limnology for fish ponds

The foregoing should demonstrate several limnological constraints to successful static water aquaculture. Fish ponds are, by necessity, highly eutrophic. Thus, the development of thermal stratification with a resultant anoxic hypolimnion and sediment surface is inevitable in the temperate zone. For this reason, fish ponds are generally shallow so that wind action is sufficient to circulate the entire pond volume (Figure 13.2). Rectangular ponds may be oriented to the prevailing winds to facilitate mechanical mixing and prevention of stratification. Ponds are usually made with flat bottoms and drains so that sediments may be exposed to air for oxidation after harvest of the fish crop. When topography makes shallow ponds impossible, the adverse effects of stratification may be negated by mechanical mixing.

Eutrophication commonly becomes so severe in ponds where fish are fed complete artificial diets that dissolved oxygen falls to lethal levels. Such oxygen depletions are caused by nighttime respiration by the huge algal community that result from feed inputs. Oxygen depletions can also occur in hypereutrophic fish ponds when plankton suddenly die on a mass scale. Emergency aeration is applied in such ponds in the early morning hours to keep the fish alive until sunlight reaches the pond surface and restores dissolved oxygen levels by photosynthesis. Aeration may need to be continued for several days in the case of a plankton die-off.

Fish ponds contain autotrophic and heterotrophic components, and the relative strength of each is controlled by fertilization strategy. Autotrophic-based food chains are augmented by fertilization with inorganic sources of nitrogen and phosphorus because phosphorus is usually the factor limiting the magnitude of primary production. Heterotrophic-based food webs are supported by the addition of organic matter such as manure or plant material. The bacteria that colonize this material are the food source for zooplankton and protozoa that eventually nourish the desired fish species.

Aerobic decomposition of organic matter depends upon an adequate supply of dissolved oxygen and a suitable temperature and pH. Microbial activity in lakes occurs at temperatures from about 5°C to 35°C, and the rate of activity increases approximately twofold for every 10°C increase in temperature. Thus, microbial decomposition essentially ceases in winter and accelerates in summer in the temperate zone. Temperature effects on the rate of organic matter decomposition are particularly important in the culture of larval fish that spawn in the spring and depend upon heterotrophic food webs to provide a supply of zooplankton to coincide in time with their requirements for forage. If spring weather is too cold or too warm, the fish may not intercept the temperature-dependent production of zooplankton.

The highest rates of microbial decomposition are supported in the pH range of 7–8. Fish pond soils are often more acidic than this, and lime application may be used to increase the effects of organic fertilization.

Bacterial biomass is about 50% carbon and 10% nitrogen. Thus, organic matter decomposition also depends upon an adequate supply of

Table 13.1 Decomposition rates of plant materials with varying levels of nitrogen

Plant genus	Percent nitrogen dry weight	Oxygen consumption mg/L/5 days
Typha	1.09	1.48
Eichhornia	1.69	1.57
Spirogyra	3.18	3.29
Pithophora	3.50	3.25
Chara	3.74	3.48
Najas	4.64	3.85
Euglena	4.66	3.53
Anabaena	9.30	5.92

Source: Almazan, G. and C. E. Boyd. 1978. *Aquaculture* 15: 75.

nitrogen. If the organic matter being decomposed contains much nitrogen, microbes will grow well and some nitrogen will be mineralized into the environment. When organic matter low in nitrogen is being decomposed, nitrogen will be immobilized from the environment in order to support microbial growth and the rate of decomposition may be nitrogen limited. Almazan (1974) provides data on the effects of nitrogen content on the rate of decomposition of some plant materials (Table 13.1).

Inorganic nitrogen may be added to nitrogen-deficient muds to increase decomposition rates. Nitrogen-rich organic fertilizers like alfalfa, seed meals, and grain brans provide excess nitrogen to decompose other materials in the pond.

Coarse organic materials, high in lignans, are more resistant to decay than softer materials. Organic fertilizers may be dried and ground to accelerate their decomposition rates.

The autotrophic and heterotrophic pathways in fertilized ponds are interrelated. Inorganic fertilization greatly increases the proliferation of organic matter in the pond that eventually dies and is decomposed by primary heterotrophs. Organic fertilizers contain small amounts of plant nutrients that increase primary production.

Fish ponds are nearly always fertilized to increase fish production, and fertilization strategies may affect the resulting communities of fish food organisms. Fertilization of ponds for the production of phytoplankton filter feeders is most effective when inorganic fertilizers high in phosphorus are used. Ponds used for the production of fish that feed on zooplankton appear to be most effectively fertilized with organic matter. Apparently, the bacteria–protozoan–zooplankton food web is more reliable than the phytoplankton–zooplankton pathway. Some workers have suggested that fertilizer nutrient levels and ratios can be manipulated to increase fish production. Desirable unicellular algae might have a selective advantage over less desirable filamentous and colonial forms when nutrient levels are low (Culver 1991). High fertilization rates of nitrogen might favor green algae over undesirable *Cyanobacteria*, which fix nitrogen, and periodic high levels of phosphorus might favor colonial green algae, which can store excess phosphorus for future use. When fertilizers are applied early in the spring and nutrient levels are maintained by frequent fertilizer applications, the shade from the resulting plankton bloom may prevent or retard the growth of macrophytes.

The highest levels of fish production occur when fish of low trophic status are selected. This is obvious in the husbandry of terrestrial animals where herbivores are always selected. Unfortunately, the most desirable fish species are often too high on the food chain for production without the use of expensive artificial diets. The lack of high-quality fish species that are also low on the food chain is an important constraint to the expansion of aquaculture in the industrialized West and helps explain the greater importance of aquaculture in the tropics and Asia where herbivorous fish are more available and more widely accepted by consumers.

References

Almazan, G. 1974. Studies on oxygen consumption by microbial organisms during decomposition of aquatic plants. MS Thesis, Auburn University, Auburn, Alabama.

Almazan, G. and C. E. Boyd. 1978. Plankton production and tilapia yields in ponds. *Aquaculture* 15: 75.

Cole, G. A. 1994. *Textbook of Limnology*. Waveland Press, Long Grove, Illinois.

Culver, D. A. 1991. Effects of the N:P ratio for fish hatchery ponds. *Verrhandlungen Internationale Vereirigung fur Theoretische und Angewandte Limnologie* 24: 1503–1507.

chapter fourteen

Principles of static water aquaculture

Static water fish culture dates back to antiquity when Chinese rice farmers discovered that fish trapped in their fields at flooding were recovered at a larger size after the fields were drained for harvest.

The most significant volume of research on the culture of fish in static water in North America began in 1934 when H. S. Swingle, an entomologist with the Alabama Agricultural Experiment Station, constructed his first pond near the Auburn University campus. Swingle, unlike contemporary fishery biologists, viewed fish culture as similar to the production of an agricultural crop. He was the first fishery biologist to apply agricultural research methods to fishery biology. His experiments replicated treatments to measure the responses of organisms to experimental treatments. The importance and scope of Swingle's contribution to the development of static water aquaculture technology is attributed to his novel approach to biological investigation. Swingle, his students, and his coworkers used the results of these experiments to develop a set of ecological principles that govern fish production. Although the bulk of Swingle's work focused on the dynamics of largemouth bass and bluegills nourished through fertilizer-enhanced natural food chains, the resulting principles are pertinent to all aspects of static water aquaculture.

Swingle's principles

Though never formally published, the following set of principles has been carried down from Swingle to his students and theirs:

1. All fish depend on plants for food.
    The first principle of general animal husbandry and life on Earth is that primary productivity—the fixation of carbon by plants and other photosynthesizers—is the base of a food chain that nourishes all trophic levels.
2. The weight of fish that can be produced depends upon the ability of the water to produce plants.
    The food chain is pyramid-shaped, so that the size of the base determines the sizes of the higher trophic level slices.

3. The ability of the water to produce plants depends upon sunlight, temperature, gases such as carbon dioxide (CO_2), and minerals (nutrients).

   The magnitude of primary production (size of pyramid base) may be limited by sunlight, the energy source for photosynthesis; temperature, which affects the rates of biological reactions; available CO_2, the carbon source for photosynthesis; or fertilizer nutrients, nitrogen and phosphorus being the most important of these.

4. The fertility of water depends upon the fertility of the watershed.

   An important limnological principle is that lakes are nutrient sinks. Fertilizer nutrients are transported to lakes and ponds by the runoff that maintains their water levels. Watersheds rich in these minerals support lakes with higher productivity more so than watersheds that are poor in fertilizer nutrients.

5. The fertility of water can be increased by adding nutrients.

   Based on the observation of Principle 4, Swingle performed many experiments on the fertilization of ponds with chemical fertilizers in order to increase fish production. The results were dramatic, and fertilization remains an essential management input for many of the world's static water aquacultures.

6. After essential nutrients have been supplied, the next limiting factor to plant production is CO_2.

   One of the most important limnological principles affecting static water aquaculture is that, unlike in terrestrial systems, the amount of CO_2 available to fuel the photosynthesis reaction is variable in aquatic systems. Primary productivity in water is often carbon limited. One of Swingle's successors at Auburn, Dr. C. E. Boyd, has conducted extensive investigations on the use of agricultural limestone to increase photosynthesis rates and, hence, fish production, in carbon-deficient waters.

7. After food, provided by nutrients and CO_2, dissolved oxygen is the next limiting factor in fish production.

   Dissolved oxygen becomes a limiting factor at the highest levels of management intensity when the plankton bloom resulting from high levels of feeding—or, more rarely, high fertilization rates—is so intense that its nighttime respiration causes an oxygen depletion sufficient to affect fish health or survival.

8. The more fertile the water is, the denser the plankton becomes, resulting in more shallow light penetration and photosynthesis.

   As pond fertility is increased through aquaculture management, the resultant plankton bloom becomes self-shading, and the depth of the zone of light penetration (photic zone) is reduced. This important limnological principle and its effect on aquaculture is addressed in Chapter 16.

9. The longer the food chain, the lower the fish production.

   In ponds with equal fertility, herbivorous fish will achieve greater production than species higher on the food chain. Eltonian ecology, predicated on the laws of thermodynamics, has critical significance to fisheries and aquaculture. Because the food chain is pyramid-shaped, significant efficiency is lost by cropping the higher trophic levels.

10. A pond has a particular carrying capacity dependent on fish species and fertility.

    The combination of Principles 2 and 9 tells us that the equilibrium weight of fish in a pond (carrying capacity) depends upon fertility and the distance of removal of the aquaculture fish species from primary productivity. Carrying capacity can thus be increased by fertilization (if photosynthesis is nutrient limited), liming (if photosynthesis is carbon limited), or by replacing the aquaculture fish species with one at a lower trophic level.

11. This carrying capacity may be a large number of small fish or a smaller number of larger fish.

    Therefore, the size of fish harvested may be regulated by the stocking rate if reproduction does not occur. One of Swingle's initial findings was that when reproduction of stocked fish occurred, the harvest was composed of large numbers of fish below his designated satisfactory harvest size. Among Swingle's best-known scientific contributions was the use of largemouth bass to control bluegill reproduction, thereby providing satisfactory fishing for both species. The following table shows how this principle can be used to control the harvest size of channel catfish (a species that does not reproduce in ponds) by regulating the stocking rate:

Stocking rate (Number per acre)	Average harvest size (One summer)
20,000	20 cm
50,000	15 cm
100,000	8 cm
200,000	5 cm

12. Where reproduction does occur, suitable fish crops depend upon a balance between predator and prey species.

    This is where we deviate from aquaculture to fish management. Swingle's work with largemouth bass and bluegills comprises a classic treatise on animal predation and a basis for the management of natural fish populations in lakes and streams. Use of a predator to control reproduction of a target species is not an important aquaculture management tool and is beyond the scope of this book.

13. The amount of food required for maintenance of a fish or fish population is less than that required for growth.

    Unlike isotherms, fish may be maintained at a constant weight indefinitely if the availability of food—either produced naturally in the pond or added as a supplement—is sufficient to meet metabolic requirements, but insufficient to provide for growth. This level of feeding is called a maintenance ration.

14. If food abundance (fertility) remains constant, the fish population will expand until the available food reaches the amount required for maintenance.

    When the available food becomes the maintenance ration, fish growth ceases and carrying capacity has been achieved. This principle will be examined in detail in Chapter 15.

15. When fish are fed, the ration must exceed the maintenance level in order for growth to occur.

    Swingle began work on the use of supplemental feeds late in his career. His successors at Auburn and other southern American universities continued this work, which led to the American channel catfish culture industry and the international intensification of the production of tilapia and other warmwater species.

16. Maximum feeding rates depend upon the ability of the aquatic system to dispose of wastes.

    Feeding provides a level of production intensity higher than that achieved by fertilization and/or liming. Fish production in fertilized ponds is limited by the amount of natural fish food provided by the fertilization regime. However, when artificial diets are added externally, carrying capacity is limited by the deterioration of water quality and its effect on fish health. Successful static water aquacultures that rely on artificial diets require water quality management to maintain satisfactory water quality for the fish.

chapter fifteen

The ecology of static water aquaculture

Fish production in standing water is normally measured in fish weight per unit of surface area per year. The world's oceans produce approximately 2 kg/Ha year, while inland lakes typically produce 10–400 kg/Ha year. Fish ponds and artificial lakes that are managed to increase fish production over what would naturally occur may produce anywhere from 100 to 4000 kg/Ha year depending upon the intensity of management inputs.

Fish production in weight per unit of surface area for a static water aquaculture is in contrast to the units of weight per unit of water flow used in flowing water aquacultures. A trout raceway 15 m × 3 m × 0.6 m (L:W:D) holding fish at a density of 80 kg/m^3 would contain a fish weight of 480,000 kg per Ha of water surface. This density is achieved by a water flow of approximately 20 L/min and would not be possible without this high water flushing rate. Thus, fish weight per unit of water surface is an inappropriate measure for flowing water aquaculture and claims of extraordinary levels of fish production should be evaluated with regard to water flow rates used to achieve them.

The potential fish production of a static water aquaculture depends not only on the intensity of management, but upon the trophic status of the selected culture species. Fish at lower trophic levels achieve higher levels of production than do those at higher trophic levels as noted by Swingle (Principle #9). For this reason, the most efficient static water aquacultures rely on herbivorous and omnivorous fish species. Zooplanktivorous species may be grown in combination with herbivores and omnivores to provide a secondary fish crop by utilizing the zooplankton food resource. Larval fish that feed on zooplankton may be reared for short culture periods in ponds fertilized to produce zooplankton, but fish production is much lower than for lower trophic level fish. Reasonable levels of production of insectivorous and piscivorous fish require high fat and protein artificial diets.

This essential ecological concept is the biological manifestation of the second law of thermodynamics, first observed by Elton (1927) and explained by Lindeman and Hutchinson (Lindeman 1942). Biological food chains are pyramid-shaped with plants at the base and carnivores at the apex because only a small fraction—roughly 10% of the energy in one trophic level—is passed on to the next; the rest is released as waste heat. This explains the relative scarcity of large carnivorous animals in the world's ecosystems and the ecological impossibility of achieving high levels of production of carnivorous fish in an aquaculture without the addition of high-energy artificial diets.

Examples from aquaculture illustrate this essential concept. Swingle spent most of his career studying the pond production of bluegill and largemouth bass. Bluegills are strict insectivores in the medium sizes and bass are strict piscivores in larger sizes. The fry of both species eat zooplankton. Large bluegills may occasionally consume small fish and medium-sized bass supplement their diets with insects. A fertilized pond might contain 300 kg of fish per Ha, but only 10%–25% of that total is occupied by bass (Swingle 1950). A pond with the same fertility would typically reach a production level of 600 kg/Ha of common carp, an omnivore, and 1500 kg/Ha of tilapia, an herbivore. The insectivorous channel catfish attains a carrying capacity of scarcely 150 kg/Ha in fertilized ponds, but production of 2500 kg/Ha is achieved when the fish are fed complete artificial diets.

Fish growth

The growth of fish in natural waters and static water fish ponds, when plotted in units of weight, generally follows an asymmetric sigmoid curve described by von Bertalanffy (1938). This growth function contains an initial exponential phase, followed by a relatively linear phase that leads to a dampened phase as fish weight reaches an asymptotic maximum (Figure 15.1).

The duration of the exponential growth stanza, the deviations of growth from exponential to linear to asymptotic phases, and the

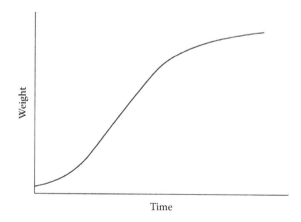

Figure 15.1 von Bertalanffy growth curve for weight gain in fish over time.

magnitude of the asymptotic maximum size vary widely among species and strains of fish (endogenous influence) and among different environmental conditions (exogenous influences). Fish species that have the potential to reach larger sizes grow more rapidly to desired harvest sizes and this relationship can be used to evaluate fish species for aquaculture (Legendre and Albaret 1991).

The exponential growth phase can be prolonged in flowing water aquaculture (Soderberg 1995) and extended in some static water aquacultures (DeSilva et al. 1991) by controlling exogenous factors affecting fish growth, but when fish pond water cannot be rapidly exchanged, one or more exogenous factors will eventually dampen fish growth.

The most important exogenous factor affecting fish growth in ponds is the quantity and quality of the available food resources. Nearly all of Earth's bodies of water support some productivity and, thus, can achieve some level of fish production. While some aquacultures rely solely on natural production, management is almost always applied to increase fish production over the level that would naturally occur. Pond production of fish is usefully described in terms of limiting factors and levels of management intensity used to overcome these limits. The first level of management intensity is fertilization to increase primary productivity. While some herbivorous fish species can grow rapidly to large sizes at the base of the food chain, the growth of most species will eventually be limited by nutritional deficiencies in fertilized ponds. The higher trophic level products of fertilization are high in fat and protein and omnivorous and carnivorous species can benefit significantly from high carbohydrate supplemental diets that spare available protein for growth. The protein provided by fertilization will eventually become limiting and continued fish growth will depend upon the feeding of increasingly complex artificial diets.

When fish in static water aquaculture are fed the nutritionally complete diets used in flowing water systems, fish growth becomes limited by the deterioration of water quality. While ammonia, nitrite, and excessively high pH levels caused by photosynthesis contribute to water quality degradation, the most important endogenous factor affecting fish growth, after food, is dissolved oxygen (DO). At the highest practical levels of management intensity, plankton, resulting from the fertilizer nutrients in the fish feed, becomes so abundant that nighttime respiration causes growth limiting or even lethal dissolved oxygen depletions.

Fish production

Fish biomass, or standing crop, is the product of individual fish size and the number of fish present and, thus, follows a graphical pattern with time similar to that observed for fish growth (Figure 15.2). Fish production is

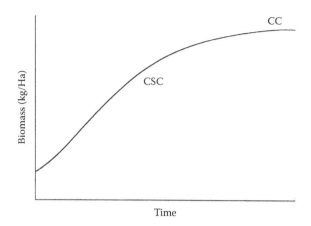

Figure 15.2 Increasing fish biomass over time in static water aquaculture. CSC = critical standing crop and CC = carrying capacity.

the biomass elaborated over time and is generally expressed in terms of kg/Ha/year.

Hepher (1978) defined the inflection point and the asymptote of this relationship in terms of food availability in a most useful ecological model to explain aquacultural fish production biological effects of different levels of management intensity.

When small fish are stocked in a pond, they are exposed to vast food resources and grow at their biological potential until food becomes limiting. The limitation in growth and, hence, production imposed by a growing fish population and a relatively constant level of food resources results in an inflection point in the production curve that Hepher (1978) called *critical standing crop* (Figure 15.2). Mathematically, it is the point where production continues to increase but at a decreasing rate. The asymptote of the production curve occurs when the available food resource in the pond is only sufficient to maintain fish body weight and the fish population ceases to increase its biomass. Hepher (1978) called this point the carrying capacity. Hickling (1971) preferred the term maximum standing stock to define the management level specific limit to static water fish production.

Hepher (1978) illustrated his theory graphically, showing that fish supplied with an excess of food grow at a species- (and temperature-) specific maximum rate (Figure 15.3). The fish growth rate begins to decrease at the point of critical standing stock when the food availability is equal to the ration required to sustain maximum fish growth.

The concept of critical standing stock should be clear to the student of flowing water fish culture in which growth and feeding rates are formalized more quantitatively than is common in static water

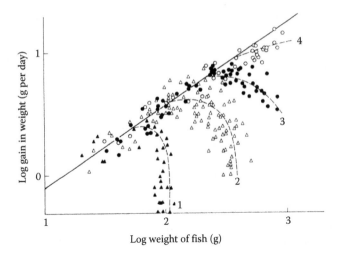

Figure 15.3 Progression of critical standing stock and carrying capacity values for common carp in increasing levels of pond management intensities. (Redrawn from Hepher, B. 1978. Ecological aspects of warmwater fish pond management. Pages 447–468 *in* S. D. Gerking, editor. *Ecology of Freshwater Fish Production.* John Wiley & Sons, New York, New York.)

aquaculture. The maximum growth rate in length increments per day is known for a given fish species-temperature combination (Soderberg 1992). This value, called ΔL, is used to precisely calculate the amount of feed required to achieve it with Haskell's (1959) equation (see Chapter 2);

$$F = 3 \times C \times \Delta L \times 100/L$$

where F = the feeding rate in percent body weight per day, C = the food conversion rate, ΔL = the daily length increment, and L = the length of the fish on the date fed. While the conventions of ΔL and F are not generally used in static water fish culture, the foregoing analysis shows that feeding above the rate of F would not increase fish growth and feeding below F would result in fish growth below ΔL and serves to explain Hepher's concepts.

Hepher (1978) showed a progression of critical standing stock and carrying capacity values for increasing levels of pond management intensities (Figure 15.3).

Fish deviate from their biological potential for growth when a limiting factor is reached. If that factor is alleviated by management, maximum growth continues until a new critical standing stock is reached. Hepher's (1978) data from common carp production experiments graphically depict

Table 15.1 Critical standing crop and carrying capacity values for carp ponds in Israel

Pond treatment	Critical standing crop (kg/Ha)	Carrying capacity (kg/Ha)
No fertilization, no feed	65	130
Fertilization, no feed	140	480
Fertilization, cereal feed	550	2500 (Estimated)
Fertilized, 25% protein pellet feed	2400	Not determined

Source: Adapted from Hepher, B. 1978. Ecological aspects of warmwater fish pond management. Pages 447–468 *in* S. D. Gerking, editor. *Ecology of Freshwater Fish Production.* John Wiley & Sons, New York, New York.

the ecology of static water aquaculture. Critical standing crops and carrying capacities for carp ponds in Israel are provided by Hepher (1978; Table 15.1).

Diana (1997) showed how critical standing stock and carrying capacity of tilapia, *O. niloticus*, differ with two different fertilization regimes (Figure 15.4) and demonstrated that tilapia at increasing levels of pond management intensity follow the pattern shown for carp by Hepher (1978). The values for critical standing stock and carrying capacity for tilapia in fertilized ponds in the tropics were estimated to be 354 and 3190 kg/Ha, respectively.

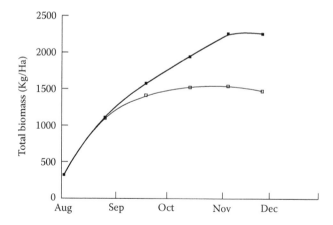

Figure 15.4 Production curves for tilapia (*O. niloticus*) under two different fertilization programs. The upper line shows carrying capacity and critical standing for a pond fertilized with organic fertilizers and the lower line shows that carrying capacity is less in a pond treated with inorganic fertilizers. (Redrawn from Diana, J. S. 1997. Feeding strategies. Pages 245–262 *in* H. S. Egna and C. E. Boyd, editors. *Dynamics of Pond Aquaculture.* CRC Press, Boca Raton, Florida.)

SAMPLE PROBLEMS

1. You stocked 15,000 25-g tilapia in a 3.5-Ha fertilized pond on April 1 and conducted sample counts throughout the growing season. These are the data you collected:

Date	Average wt. (g)
April 1	25
April 21	74
May 10	126
June 1	180
June 17	230
July 9	260
Aug 5	305
Aug 20	310
Oct 1	330
Oct 20	325

   Assuming that there was no mortality or reproduction, what is the carrying capacity of the pond? What is the critical standing stock? If the fish had been fed, on what date would they have reached 500 g? If the pond would have been stocked with 25,000 fish, how large would they have been on November 1? If 10,000 had been stocked, how large would they have been on November 1? If 250 kg/Ha were harvested on July 1 and the remainder of the fish were harvested on November 1, what would be the annual production of the pond?

2. You stocked 4000 mirror carp in a pond on the first day in spring on which the water temperature reached 15°C. At the end of the summer, the fish weighed 350 g each. How many would you stock next year if you want their average weight to be 500 g at the end of the summer?

3. Fish stocked in a fertile pond at 4000/Ha and 100 g each grew at a rate of 1% of their body weight per day for 50 days until the growth rate began to decline. The rate of decline in growth rate was 0.01% (growth rate was 0.9% of body weight on day 51, 0.8% of body weight on day 52, etc.). If feeding had been started on day 50, the maximum rate of growth would have continued for another 50 days before it began to decline. What are the critical standing stocks and carrying capacities for the unfed and fed fish in this pond?

4. Consider a 1-Ha pond 2 m deep with a photic zone averaging 1 m deep. The photosynthesis rate in the photic zone is 5 mg CO_2 per L per day. During a 180-day growing season, 200 kg of fish are produced. If fish are 75% moisture and 50% of their dry matter is carbon,

what percentage of the carbon fixed in photosynthesis ended up in fish biomass?

5. 2000 kg/Ha of fish are harvested from a pond every year. There are no nitrogen compounds in the water supply, no nitrogen is available from the pond mud, and the fish are not fed. Thus, all nitrogen contained in the harvested fish must come from the planktonic fixation of nitrogen gas. If the fish are 75% water and 75% protein on a dry-weight basis, what must be the minimum nitrogen fixation rate in the pond in kg of N/Ha/year?

References

DeSilva, S. S., R. M. Gunasekera, and K. F. Shim. 1991. Interactions of varying dietary protein and lipid levels in young red tilapias: Evidence of protein sparing. *Aquaculture* 95: 305–318.

Diana, J. S. 1997. Feeding strategies. Pages 245–262 *in* H. S. Egna and C. E. Boyd, editors. *Dynamics of Pond Aquaculture*. CRC Press, Boca Raton, Florida.

Elton, C. E. 1927. *Animal Ecology*. Macmillan Publishers. New York, New York.

Haskell, D. C. 1959. Trout growth in hatcheries. *New York Fish and Game Journal* 6: 205–237.

Hepher, B. 1978. Ecological aspects of warmwater fish pond management. Pages 447–468 *in* S. D. Gerking, editor. *Ecology of Freshwater Fish Production*. John Wiley & Sons, New York, New York.

Hickling, C. F. 1971. *Fish Culture*. Faber and Faber. London, UK.

Legendre, M. and J. J. Albaret. 1991. Maximum observed length as an indicator of growth rate in tropical fishes. *Aquaculture* 94: 327–341.

Lindeman, R. L. 1942. The trophic dynamic aspects of ecology. *Ecology* 23: 399–418.

Soderberg, R. W. 1992. Linear fish growth models for intensive aquaculture. *Progressive Fish-Culturist* 54: 255–258.

Soderberg, R. W. 1995. *Flowing Water Fish Culture*. Lewis Publishers, Boca Raton, Florida.

Swingle, H. S. 1950. Relationships and dynamics of balanced and unbalanced fish populations. Alabama Agricultural Experiment Station, Alabama Polytechnic Institution, Auburn, Alabama.

von Bertalanffy, L. 1938. A quantitative theory of organic growth. *Human Biology* 10: 181–213.

chapter sixteen

Photosynthesis and respiration

Concentrations of dissolved oxygen (DO) and carbon dioxide (CO_2) in water fluctuate in a dynamic diel cycle following the processes of oxygenic photosynthesis,

$$CO_2 + H_2O = CHO \, (\text{Carbohydrate}) + O_2$$

and aerobic respiration,

$$CHO + O_2 = CO_2 + H_2O$$

The oxygen-producing, light-dependent reactions of oxygenic photosynthesis occur in the daylight hours and at water depths receiving sufficient light to power the reaction, or until the energy and reducing power for fixation of CO_2 to carbohydrates is exhausted. Aerobic respiration occurs at all times and depths that DO is present. In the absence of DO, carbohydrate continues to be metabolized through the process of anaerobic respiration or fermentation. The general expression for fermentation is

$$CHO = CH\text{–}OH \, (\text{alcohol or organic acids}) + CO_2$$

The products resulting from fermentation or anaerobic respiration are further reduced under extended periods of anoxia to methane (CH_4) and other longer chain compounds containing only carbon and hydrogen. Long periods of anoxic decomposition have produced the vast fields of fossil fuels upon which our society depends for its energy needs.

Unlike on land, where atmospheric gas concentrations remain constant regardless of the levels of photosynthesis and respiration, DO and CO_2 in fish ponds fluctuate greatly and may be depleted and reach supersaturated levels in a single 24-hour cycle. More detailed explanations of aquatic photosynthesis include Falkowski and Raven (1977) and Riebesell and Wolf-Gladrow (2002).

Solubility of DO and CO_2

Henry's law describes the solution of atmospheric gases in water. The equilibrium concentration of a gas in water is related to its partial pressure in the atmosphere. The partial pressure in the atmosphere is the fraction of the total atmospheric pressure of that gas according to Dalton's law. Procedures for determining the solubility of oxygen in water at any temperature, atmospheric pressure, and salinity are provided in Chapter 5. The solubility of CO_2 is determined similarly using the appropriate Bunsen coefficients. The partial pressure of CO_2, whose volume fraction in the atmosphere has risen to 0.040% (2015) (Table 5.1), is $760 \times 0.00040 = 0.30$ mm Hg.

The Bunsen coefficient for CO_2 at 20°C is 0.8705 (Table 16.1). Thus, the solubility of CO_2 at 1.0 atm of pressure and a temperature of 20°C is

$$0.8705 \times \frac{44,000\,mg}{mole} \times \frac{mole}{24\,L} \times 0.00040 = 0.64\,mg/L$$

The following polynomial was developed from the salinity and temperature effects on CO_2 solubility presented by Colt (1984).

$$Ds_{CO_2} = 0.062 - 0.004T + 0.00004T^2 + 0.002S$$

where Ds_{CO_2} = decrease in CO_2 caused by salinity in mg/L and S = salinity on parts per thousand (‰).

Measurement of dissolved gas concentrations

Dissolved gas concentrations fluctuate in static water fish ponds due principally to the daily cycle of photosynthesis and respiration. Thus water used for aquaculture seldom contains the equilibrium concentrations, and the actual gas concentrations must be measured. Measurement of DO levels in water is discussed in Chapter 5. Carbon dioxide concentrations are measured by titration with standard base to the phenolthalein end point (Boyd and Tucker 1992). Carbon dioxide levels can also be determined indirectly from the equilibrium expression by measuring alkalinity and pH (see Chapter 17). However, because CO_2 behaves poorly as an ideal gas and the dynamics of pH and CO_2 are so rapid in fish ponds, the value determined from the equilibrium expression is not likely to reflect the actual CO_2 level.

CO_2 and pH dynamics in fish ponds

Recall from Chapter 5 that nearly all CO_2 in water comes from the dissolution of earth carbonates that ultimately results in free CO_2 for aquatic photosynthesis. Calcite precipitates at about pH 8.3 due to the initiation of production of CO_3^{2-} from HCO_3^-. Thus the increase in pH caused by the photosynthetic removal of CO_2 should be checked at pH 8.3.

Table 16.1 Bunsen coefficients and air solubility for carbon dioxide in moist air at 1.0 atm of pressure

Temperature	Bunsen coefficient	Solubility (mg/L)
0	1.7272	1.29
1	1.6604	1.23
2	1.5972	1.18
3	1.5373	1.13
4	1.4805	1.09
5	1.4265	1.05
6	1.3753	1.00
7	1.3267	0.97
8	1.2805	0.93
9	1.2365	0.89
10	1.1947	0.86
11	1.1548	0.83
12	1.1169	0.70
13	1.0808	0.77
14	1.0464	0.74
15	1.0136	0.72
16	0.9822	0.69
17	0.9523	0.67
18	0.9238	0.65
19	0.8965	0.63
20	0.8705	0.61
21	0.8455	0.59
22	0.8217	0.57
23	0.7989	0.55
24	0.7771	0.53
25	0.7562	0.52
26	0.7362	0.50
27	0.7170	0.49
28	0.6986	0.47
29	0.6810	0.46
30	0.6641	0.45
31	0.6478	0.43
32	0.6323	0.42
33	0.6173	0.41
34	0.6029	0.40
35	0.5891	0.39
36	0.5759	0.39

(Continued)

Table 16.1 (Continued) Bunsen coefficients and air
solubility for carbon dioxide in moist air at 1.0 atm
of pressure

Temperature	Bunsen coefficient	Solubility (mg/L)
37	0.5631	0.37
38	0.5509	0.36
39	0.5391	0.35
40	0.5277	0.34

Source: Bunsen coefficients are from Colt, J. E. 1984. *Computation of
Dissolved Gas Concentrations in Water as Functions of
Temperature, Salinity, and Pressure.* American Fisheries Society
Special Publication 14. Bethesda, Maryland.

The speciation of inorganic carbon in water is pH controlled, but in
actuality this system in aquatic environments has little chance of achieving
equilibrium. There is a theoretical tendency for CO_2 in solution to equilibrate
with atmospheric CO_2 and for CO_3^{2-} to equilibrate with the solid phase,
but this is not achieved due to the slowness of the conversions among CO_2,
HCO_3^-, and CO_3^{2-}. For this reason, eutrophic aquatic systems like fertilized fish ponds may experience such nonequilibrium phenomena as under
or super saturation of CO_2 and increases in pH to dangerously high levels.

We have seen that free CO_2 is rarely a primary component of dissolved inorganic carbon. This should cause aquatic photosynthesis to be
carbon limited at higher pH levels due to the slow conversion of HCO_3^-
to CO_2. This deficiency is countered by the ability of some algae to utilize
HCO_3^- as a carbon source. A few species of algae require HCO_3^- and
cannot grow on CO_2. When aquatic plants have similar affinities for CO_2
and HCO_3^-, utilization of HCO_3^- occurs when its level exceeds that of
CO_2 by more than about 10 times. This is the case in many freshwaters.
When pH exceeds 8.3, CO_2 is inadequate to supply photosynthesis and
HCO_3^- utilization by algae so capable causes pH elevation in low alkalinity waters. The sea is a saturated solution of $CaCO_3$ and thus has a pH
of about 8.3. High levels of primary production in marine environments
can be explained by HCO_3^- utilization by marine algae. Bicarbonate
utilization occurs through active HCO_3^- uptake and/or extracellular conversion of HCO_3^- to CO_2 by the enzyme carbonic anhydrase. This may
cause an increase in pH at the diffusive boundary layer surrounding the
alga which decreases CO_2 and increases CO_3^{2-}. In an attempt to maintain
equilibrium, CO_2 uptake is followed by spontaneous or carbonic anhydrase catalyzed conversion of HCO_3^- to CO_2 at the cell surface. Thus,
aquatic photosynthesis continues at high pH levels and contributes to the
rise in pH to nonequilibrium levels (>8.3) and exacerbates problems of
high pH caused by fertilization in fish ponds.

High pH levels affect fish directly. Furthermore, high pH values in the afternoon favor the fraction of NH_3 in an ammonia solution. Thus, NH_3 can reach extreme levels for brief periods in the daily pH cycle. High pond pH values (>9.0) can result in fish mortality, either from the toxicity of pH, the resulting high levels of un-ionized ammonia (NH_3), or a combination of these two effects.

Afternoon pH values as high as 10.0 were recorded in fertilized ponds used to culture walleye, and pH and fish survival were positively correlated (Soderberg and Marcinko 1999). Soderberg (1985) and Soderberg et al. (1983) reported a relationship between the average daily exposure to ammonia and the growth and survival of rainbow trout grown in static water ponds. Channel catfish growth and survival were not affected by high afternoon NH_3 exposures, but gill lesions characteristic of chronic ammonia exposure were more prevalent in fish in ponds with higher ammonia levels (Soderberg et al. 1984).

Aqueous ammonia exists in two molecular forms, NH_3 and NH_4^+, and their combined concentration is referred to as TAN (total ammonia nitrogen). Un-ionized ammonia (NH_3) is toxic to fish, and the equilibrium between the two forms is dependent on pH and temperature. Instructions for calculating the un-ionized fraction in a TAN solution are provided in Chapter 9, but for most static water aquaculture applications, the values provided in Table 16.2 will suffice. Note that for a given temperature, the exposure of fish to NH_3 increases dramatically with pH.

Ball (1967) reported that the 2-day LC_{50} (concentration resulting in 50% mortality) of NH_3 to rainbow trout is 0.41 mg/L. Colt and Tchobanoglous

Table 16.2 Fraction of NH_3 in an ammonia solution

| Temperature (°C) | pH | | | | |
	6.5	7.0	7.5	8.0	8.5
20	0.0013	0.0039	0.0124	0.0381	0.1112
21	0.0013	0.0042	0.0133	0.0408	0.1186
22	0.0015	0.0046	0.0143	0.0438	0.1264
23	0.0016	0.0049	0.0153	0.0469	0.1356
24	0.0017	0.0053	0.0164	0.0502	0.1431
25	0.0018	0.0056	0.0176	0.0537	0.1521
26	0.0019	0.0060	0.0189	0.0574	0.1614
27	0.0021	0.0065	0.0202	0.0613	0.1711
28	0.0022	0.0069	0.0216	0.0654	0.1812
29	0.0024	0.0074	0.0232	0.0697	0.1916
30	0.0025	0.0080	0.0248	0.0743	0.2025
31	0.0027	0.0085	0.0265	0.0791	0.2137
32	0.0029	0.0091	0.0283	0.0842	0.2253

(1978) reported the LC_{50} of NH_3 to channel catfish is 2–3.1 mg/L. The 72-hour LC_{50} of NH_3 to tilapia (*Oreochromis aureus*) has been reported at 2.35 mg/L (Redner and Stickney 1979). Rainbow trout and channel catfish exhibit reduced growth at NH_3 concentrations of 0.016 mg/L (Larmoyeaux and Piper 1973) and 0.048 mg/L (Colt and Tchobanoglous 1978), respectively.

SAMPLE PROBLEMS

1. Calculate the solubilities of DO and CO_2 in a 29°C catfish pond at 200 m above sea level.
2. Calculate the solubilities of DO and CO_2 in a 28°C shrimp pond at sea level with a salinity of 17%.
3. Explain how plankton blooms can occur in the ocean when the chart shows that there is no CO_2 at the pH of seawater.
4. A tilapia pond in Thailand contains 1 mg/L of TAN. The temperature is 28°C. The pH in the morning is 6, and in the afternoon it is 9. How much ammonia are the fish in this pond exposed to? Is this a dangerous exposure?
5. A walleye pond in Pennsylvania, temperature 20°C, TAN, 0.3 mg/L, experiences a dusk pH of 10. Could this have killed the fish?

References

Ball, I. R. 1967. The relative susceptibilities of some species of freshwater fish to poisons—I. Ammonia. *Water Research* 1: 767–775.

Boyd, C. E. and C. S. Tucker. 1992. *Water Quality and Pond Soil Analysis for Aquaculture*. Alabama Agricultural Experiment Station, Auburn University, Alabama.

Colt, J. E. 1984. *Computation of Dissolved Gas Concentrations in Water as Functions of Temperature, Salinity, and Pressure*. American Fisheries Society Special Publication 14. Bethesda, Maryland.

Colt, J. and G. Tchobanoglous. 1978. Chronic exposure of channel catfish, *Ictalurus punctatus*, to ammonia: Effects on growth and survival. *Aquaculture* 15: 353–372.

Falkowski, P. G. and J. A. Raven. 1977. *Aquatic Photosynthesis*. Blackwell Science, Oxford, England.

Larmoyeaux, J. C. and R. G. Piper. 1973. Effects of water reuse on rainbow trout in hatcheries. *Progressive Fish-Culturist* 35: 2–8.

Redner, B. D. and R. R. Stickney. 1979. Acclimation to ammonia by *Tilapia aurea*. *Transactions of the American Fisheries Society* 108: 383–388.

Riebesell, U. and D. A. Wolf-Gladrow. 2002. Supply and uptake of inorganic nutrients. Pages 109–140 *in* P. J. le B. Williams, D. N. Thomas, and C. S. Reynolds, editors, *Phytoplankton Productivity*. Blackwell Sciences, Oxford, England.

Soderberg, R. W. 1985. Histopathology of rainbow trout, *Salmo gairdneri* Richardson, exposed to diurnally fluctuating un-ionized ammonia levels in static-water ponds. *Journal of Fish Diseases* 8: 57–64.

Soderberg, R. W., J. B. Flynn, and H. R. Schmittou. 1983. Effects of ammonia on growth and survival of rainbow trout in intensive static-water culture. *Transactions of the American Fisheries Society* 112: 448–451.

Soderberg, R. W. and M. T. Marcinko. 1999. Substitution of granular for liquid fertilizers for the pond production of walleye in earthern ponds. *Journal of Applied Aquaculture* 9: 33–44.

Soderberg, R. W., M. V. McGee, and C. E. Boyd. 1984. Histology of cultured channel catfish, *Ictalurus punctatus* (Rafinesque). *Journal of Fish Biology* 24: 683–690.

chapter seventeen

The carbon cycle, alkalinity, and liming

Primary production on land is fueled by a constant (400 ppm in 2015) supply of CO_2. One of the most important principles of limnology, however, is that dissolved CO_2 levels in water are variable. Not only does the supply of CO_2 vary greatly during the diurnal cycle as it is depleted by photosynthesis and replenished by respiration, but the total amount of CO_2 available to participate in the diurnal cycle varies from water body to water body. This variability is of fundamental concern to the study of static water fish culture because low availability of CO_2 in water may limit primary productivity and hence, fish production. Management in the form of addition of agricultural limestone is then needed to supply the necessary CO_2 to fuel photosynthesis at a satisfactory level.

Unlike in the case of dissolved oxygen in water, the principal source of CO_2 for aquatic photosynthesis is not the atmosphere. Carbon dioxide is dissolved in water according to Henry's law, but this source is a small fraction of the amount required to explain the productivity that we observe in lakes, rivers, the sea, and fish ponds. In fact, if atmospheric dissolution were the only source of CO_2 in water, there would be no limnology and no fish culture because photosynthetic plant production would be insufficient to support an aquatic consumer like a fish.

Carbon dioxide for aquatic photosynthesis is supplied primarily from the dissolution of earth carbonate rocks. The principle earth carbonate is calcite ($CaCO_3$), but other minerals such as dolomite ($CaMg(CO_3)_2$), collectively called limestone, also release CO_2 as they dissolve.

Carbon cycle

Carbon, like water, participates in a global cycle. Only a trace participates in photosynthesis on land and in the photic zones of aquatic and marine environments, while the vast majority is tied up in earth carbonate rocks (Table 17.1).

Carbonate rocks are formed by marine organisms capable of precipitating $CaCO_3$ from seawater to make their shells or skeletons. The most obvious of these processes is that conducted by corals. Coral reefs are formed by corals, skeleton-secreting colonial animals that grow in

Table 17.1 Global carbon budget

Component	Amount × 10⁹ metric tonnes	Percent of total
Atmosphere	710	0.0035
Biomass on land	590	0.0029
Litter on land	60	0.0003
Soil	1670	0.0083
Fossil fuels	5000	0.0250
Earth carbonates	20,000,000	99.76
Ocean biomass	4	0.00002
Dissolved in seawater		
Surface	680	0.0034
Intermediate depths	8200	0.0409
Deep waters	26,000	0.1297
Ocean sediments	4900	0.0244

Source: Data from U.S. Department of Energy 1980. *Global Carbon Budget. Earth System Science Data.* Oak Ridge National Laboratory, Oak Ridge, Tennessee.

clusters of connected individuals called polyps. Each polyp secretes a cup of $CaCO_3$ and adjacent cups fuse to form the reef as carbonate sediments are trapped within the porous framework of the coral, filling some of the voids.

Carbonate platforms are structures formed by the deposition of carbonate sediments, principally composed of the skeletons of calcarious algae. Carbonate sediments also occur in deeper waters as a result of deposition of the skeletons of amoeba-like organisms called planktonic foraminifers.

Note that only 0.0244% of global carbon is presently contained in these sediments (Table 17.1). Hundreds of millions of years of tectonic rearrangement of the Earth's crust have thrust nearly all of the accumulated carbonate sediments onto land where they occur as prominent limestone formations (Figure 17.1).

When earth carbonates become liberated from the sea, they are able to continue on their path in the carbon cycle because these minerals dissolve in freshwater according to the following general reaction:

$$CaCO_3 + H_2O + CO_2 = Ca^{2+} + 2HCO_3^- \tag{17.1}$$

Bicarbonate (HCO_3^-), is related to CO_2 in water as follows:

$$CO_2 + H_2O = H^+ + HCO_3^- \tag{17.2}$$

Figure 17.1 The Santa Elena Canyon in Big Bend National Park on the Rio Grande River between Mexico and Texas is composed of solid limestone, laid down by corals in an ancient tropical sea.

The equilibrium expression for this reaction is

$$\frac{[H^+][HCO_3^-]}{[CO_2]} = 4.47 \times 10^{-7} \tag{17.3}$$

Bicarbonate further dissociates to carbonate (CO_3^{2-}):

$$HCO_3^- = H^+ + CO_3^{2-} \tag{17.4}$$

And the equilibrium expression is

$$\frac{[CO_3^{2-}][H^+]}{\left[HCO_3^-\right]} = 4.68 \times 10^{-11} \tag{17.5}$$

The way in which earth carbonates control CO_2 levels in water is summarized as follows:

$$\text{Marine Carbonate Sediments} \rightarrow \text{Earth Carbonates} + CO_2 = HCO_3^- \leftrightarrow CO_2 \rightarrow \text{Plants}$$

Note from Equation 17.1 that CO_2 is required for the dissolution of the solid phase so that two moles of HCO_3^- are released, one from the earth carbonate rock and the other from the CO_2 already present. Thus, the feedback arrow shown above is applicable.

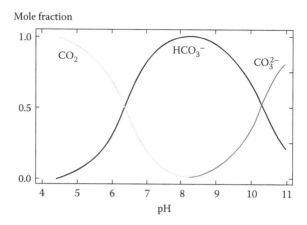

Figure 17.2 Speciation of carbon in water as a function of pH. (Redrawn from Boyd, C. E. 1990. *Water Quality in Ponds for Aquaculture.* Alabama Agricultural Experiment Station, Auburn University, Alabama.)

Because the dissociation of carbon is an H^+-liberating process, carbon speciation in water is controlled by pH as shown in Figure 17.2.

Note that HCO_3^- begins to appear in solution at a pH of 4.5. At a pH of 6.35, half of the carbon in solution is CO_2 and half is HCO_3^-. Carbonate appears in solution at a pH of 8.34, at which point CO_2 is not detectable.

pH

The negative logarithm of the molar activity of the hydrogen ion (H^+) is the pH. For example, $[H^+]$ of a solution with a pH of 7 is 1×10^{-7}. If $[H^+]$ is 1.2×10^{-6}, the pH is 5.92. The equilibrium constant for the dissociation of water is 1×10^{-14}. Thus the sum of the pH and the pOH is 14. Water with a pH of 7 is referred to as neutral because pH = pOH. The pH of a solution is measured with a pH meter that translates the current of a galvanic cell into a pH reading on a dial or a digital readout. The pH meter must be carefully calibrated with buffers of known pH.

Alkalinity

The term alkalinity refers to the concentration of titratable bases in water. These bases include HCO_3^-, CO_3^{2-}, and OH^-. Bicarbonate is predominant over CO_3^{2-} in the pH range of most natural waters (Figure 17.2) and OH^- is present in minute quantities. Therefore, alkalinity can be understood to be the concentration of HCO_3^- and CO_3^{2-}, the dissolution products of earth carbonates. Alkalinity is expressed as mg/L of $CaCO_3$, although the definition does not include the cation Ca^{2+}. Hardness is defined as the concentration of divalent cations in water, which for practical purposes are Ca^{2+} and Mg^{2+}. Hardness is also expressed as mg/L $CaCO_3$, so waters whose mineral content comes only from the dissolution of earth carbonates will have equal values for hardness and alkalinity. In fact, most natural and marine waters have similar values of hardness and alkalinity. Exceptions include alkaline lakes where Ca^{2+} is replaced by Na^+ as the cation balancing alkalinity and waters high in gypsum ($CaSO_4$). In alkaline lakes, the alkalinity is high and the hardness is low because Na^+ is not a divalent cation. In gypsum lakes, hardness is high and alkalinity is low because sulfate is not a titratable base.

The relationships between pH, alkalinity, and CO_2 may be easily determined from the equilibrium expressions. For example, we may estimate the alkalinity of a water sample that has a pH of 7.0 and a CO_2 concentration of 0.5 mg/L as follows:

$$CO_2: \frac{0.5\,mg}{L} \times \frac{mole}{44,000\,mg} = 1.136 \times 10^{-5}\,M$$

$$pH = 1 \times 10^{-7}$$

$$\frac{[H^+][HCO_3^-]}{[CO_2]} = 4.47 \times 10^{-7}$$

$$\frac{[10^{-7}][HCO_3^-]}{[1.136 \times 10^{-5}]} = 4.47 \times 10^{-7}$$

$$HCO_3^- = \frac{5.078 \times 10^{-5}\,moles}{L} \times \frac{61,000\,mg}{mole} \frac{3.097\,mg\,HCO_3^-}{L}$$
$$\times \frac{100\,(formula\ weight\ of\ CaCO_3)}{61\,(formula\ weight\ of\ HCO_3^-)}$$
$$= 5.08\,mg/L\ alkalinity$$

Suppose we wish to estimate the pH of rainwater at 10°C. Rainwater is distilled because all the minerals were left behind when the water evaporated. Thus the H^+ and HCO_3^- come only from the CO_2 dissolved in the rain as it falls to Earth, so they have equal molar concentrations. The solubility of CO_2 in water at 10°C and 1 atmosphere of pressure is 0.75 mg/L.

$$CO_2: \frac{0.75\,mg}{L} \times \frac{mole}{44,000\,mg} = 1.705 \times 10^{-5}\,M$$

$$H^+ = HCO_3^-$$

$$\frac{[H^+][HCO_3^-]}{[CO_2]} = 4.47 \times 10^{-7}$$

$$\frac{[H^+]^2}{[1.705 \times 10^{-5}]} = 4.47 \times 10^{-7}$$

$$H^+ = 2.76 \times 10^{-6}$$

pH = 5.56. The pH of pure water is 5.56. The acidity is due to dissolved CO_2.

Measurement of alkalinity

Alkalinity in water is measured by titration of a 100 mL sample with a 0.01 M solution of H_2SO_4 to a pH of 4.5, indicated by a color change from yellow to orange with methyl orange indicator. The total alkalinity is equal to the mL of titrant multiplied by 10 (Boyd and Tucker 1992).

Identification of ponds needing lime

The alkalinity at which aquatic photosynthesis is carbon-limited has been widely studied. Moyle (1946) found that hard waters are generally more productive than soft waters. Waters with total alkalinities of less than 20 mg/L $CaCO_3$ do not normally respond to fertilization (Zeller and Montgomery 1957; Thomaston and Zeller 1961; Boyd 1990). At Mansfield University we have studied the populations of largemouth bass (*Micropterus salmoides*) and sunfish (*Lepomis* sp.) in farm ponds for many years. These populations normally become structured with either the sunfish or the bass having excellent growth and the other species exhibiting stunted growth. In 50 populations with normal population structures, total alkalinities were greater than 20 mg/L $CaCO_3$. In six populations where neither of the two species grew satisfactorily, total alkalinities ranged from 18 to 23 mg/L $CaCO_3$. This is because plant production is carbon-limited at low bicarbonate concentrations, and addition of lime to ponds with total alkalinities greater than about 20 mg/L is generally ineffective (Boyd 1990). When the alkalinities of softwater ponds are increased by the addition of agricultural limestone, plant production (and hence fish production) increase due to increased levels of CO_2 for photosynthesis and greater availability of soil phosphorus caused by the elevated soil pH (Boyd and Scarsbrook 1974; Arce and Boyd 1975; Boyd 1990).

Liming

The rates of primary production in ponds that have alkalinities less than 20 mg/L may be increased by the addition of agricultural limestone. The effects of liming are accomplished by the replacement of acid cations, principally Al^{3+}, with the basic cations Ca^{2+} or Mg^{2+}. Boyd (1982, 1990) provides detailed accounts of soil chemistry effects of limestone addition to pond soils. After much experimentation, Pillai and Boyd (1985) described a procedure to determine the lime requirements of acid pond soils. A *p*-nitrophenol buffer is prepared by dissolving 10 g *p*-nitrophenol, 7.5 g H_3BO_3, 37 g KCl, and 5.25 g of KOH in 1000 mL of distilled water. The pH meter is set to read pH 8.0 with the buffer. A composite soil sample is collected from the pond bottom, dried, and pulverized. A sample of 20 g of soil is added to 40 mL of the *p*-nitrophenol buffer in a 100-mL beaker, and the pH of the resulting mixture is taken. The acidity in the soil will reduce the pH so that the pH change multiplied by 5600 is equal to the lime requirement in kg $CaCO_3$/Ha. The pH change of the buffer responds to lime requirement down to a pH of 6.8. If 20 g of soil reduces the buffer pH below 6.8, the procedure should be repeated with 10 g of soil. A complete description of the lime requirement procedure, including dealing with acid sulfate soils, is given in Boyd and Tucker (1992).

Basic materials other than $CaCO_3$ may also be utilized to neutralize soil acidity and increase alkalinity. Calcium oxide (CaO) has a neutralizing capability which is 179% that of $CaCO_3$. Equivalent values for calcium hydroxide ($Ca(OH)_2$) and dolomite ($CaMg(CO_3)_2$) are 136% and 109%, respectively.

SAMPLE PROBLEMS

1. What is the pH of a solution that contains 100 mg/L HCO_3^- and 0.5 mg/L CO_2?
2. What is the alkalinity of the solution described in Problem 1?
3. A solution contains 5 mg/L CO_2 and has a pH of 8.2. What is the bicarbonate concentration? What is the alkalinity?
4. If the lime requirement of your pond is 2116 kg of $CaCO_3$, how much quicklime (CaO) would you use? How much dolomite?
5. If you add 2000 kg of $CaCO_3$ to a 1 Ha pond which is 1 m deep and the alkalinity of the water rises from 5 to 10 mg/L, how much lime was required to satisfy the mud demand?
6. How much CO_3 is present in a water sample with a pH of 6 and a HCO_3^- concentration of 100 mg/L?
7. How much CO_3^{2-} is present in a solution with a pH of 10 and a HCO_3^- concentration of 100 mg/L?

8. Suppose a water sample has an alkalinity of 100 mg/L $CaCO_3$ and a CO_2 concentration of 10 mg/L. If half of the CO_2 is removed by photosynthesis, what will be the new pH? If three-quarters of the CO_2 is removed what will be the new pH?

9. If you add 15 mg of gypsum ($CaSO_4$) to 1 liter of water with a hardness of 50 mg/L $CaCO_3$, what will the hardness be when all of the calcium sulfate has dissolved? How much will the alkalinity increase?

10. What is the lime requirement of a pond, 20 g of whose dried mud decreases the pH of 20 mL of *p*-nitrophenol buffer from 8.0 to 7.2?

References

Arce, R. G. and C. E. Boyd. 1975. Effects of agricultural limestone on water chemistry, phytoplankton productivity and fish production in soft-water ponds. *Transactions of the American Fisheries Society* 104: 308–312.

Boyd, C. E. 1982. *Water Quality Management for Pond Fish Culture*. Elsevier Scientific Publishing Company, Amsterdam, The Netherlands.

Boyd, C. E. 1990. *Water Quality in Ponds for Aquaculture*. Alabama Agricultural Experiment Station, Auburn University, Alabama.

Boyd, C. E. and E. Scarsbrook. 1974. Effects of agricultural limestone on phytoplankton communities of fish ponds. *Arch Hydrobiol* 74: 336–349.

Boyd, C. E. and C. S. Tucker. 1992. *Water Quality and Pond Soil Analysis for Aquaculture*. Alabama Agricultural Experiment Station, Auburn University, Alabama.

Moyle, J. B. 1946. Some indices of lake productivity. *Transactions of the American Fisheries Society* 76: 322–334.

Pillai, V. K. and C. E. Boyd. 1985. A simple method for calculating liming rates for fish ponds. *Aquaculture* 46: 157–162.

Thomaston, W. W. and H. D. Zeller. 1961. Results of a six-year investigation of chemical soil and water analysis and lime treatment in Georgia fish ponds. *Proceedings of the Annual Conference of the Southeastern Association of Game and Fish Commissioners* 15: 236–245.

U.S. Department of Energy 1980. *Global Carbon Budget. Earth System Science Data*. Oak Ridge National Laboratory, Oak Ridge, Tennessee.

Zeller, H. D. and A. B. Montgomery. 1957. Preliminary investigations of chemical soil and water relationships and lime treatment of soft water in Georgia farm ponds. *Proceedings of the Annual Conference of the Southeastern Association of Game and Fish Commissioners* 11: 71–76.

chapter eighteen

Nutrient cycles and fertilization

Static water aquaculture is the practice of applied limnology. The limnological principles of production and aquatic ecology are used to artificially augment natural production in the aquatic system of the fish pond through aquaculture management. The traditional management strategy involves the addition of organic or inorganic material, or a combination of the two, to increase productivity, which ultimately results in increased fish production. As long as photosynthesis is not carbon limited, as indicated by low levels of alkalinity, the addition of fertilizer nutrients will increase fish production until it is limited by low dissolved oxygen levels caused by excessive plankton respiration.

Aquatic food chains transform energy into fish biomass by two basic mechanisms. Autotrophic production is the direct proliferation of photosynthate by algae, *Cyanobacteria*, and photosynthetic protozoans. Heterotrophic production is the release of nutrients from organic matter through the microbial loop, accomplished primarily by bacteria. Both processes are integral to the function of aquatic systems. Autotrophic production is stimulated by the addition of inorganic fertilizers, which are manufactured products with high levels of soluble nitrogen (N) and phosphorus (P). Heterotrophic production is stimulated by the addition of organic matter, such as plant material or manure, produced directly or indirectly by photosynthesis that occurs outside of the system to which it is added.

Solar energy is transformed directly to photosynthate biomass, which may be consumed directly by fish or made into natural foods for fish by secondary production. The autotrophic energy pathway involves the production of photosynthetic microbes which are consumed by zooplankton and protozoa, which are in turn consumed by fish or invertebrates. The energy originally captured in photosynthesis is ultimately transferred to carnivores in the aquatic system.

The energy contained in organic fertilizers is released through microbial decomposition. The resulting microbial biomass and the particulate organic matter that they produce are consumed by protozoa, which are in turn consumed by zooplankton and ultimately turned into fish biomass.

As organic material is decomposed, small amounts of N and P are released that result in an autotrophic component to the production of organically fertilized fish ponds. Autotrophic and heterotrophic production both result in unconsumed biomass at all trophic levels, which is decomposed through the heterotrophic pathway, adding a heterotrophic component to the production of organically fertilized fish ponds.

Ecology of cultured fish

Fish are produced in fertilized ponds at all aquatic trophic levels. Some tilapias, silver carp, and milkfish are phytoplanktivores that feed directly on primary productivity. Mullets, crayfish, and shrimp are detritivores that consume decomposing organic matter and its attendant microbial fauna. Some tilapias and grass carp consume macrophytic vegetation, including macrophytic algae. Paddlefish, bighead carp, and larval game fish, such as walleye and striped bass, feed on zooplankton. Common carp, goldfish, mud carp, and mrigal are benthic omnivores that depend upon detritus, some living plant material, and the microbes, algae, and macroinvertebrates associated with the benthos. Black carp and some species of sunfish are moluscavores. Other sunfish species are insectivores that are reared with the piscivorous largemouth bass.

There is considerable niche overlap among most cultured fish. For instance, milkfish supplement their principal diet of phytoplankton with detritus and some macrophytic algae. Larval walleye consume some species of aquatic insect larva when available. Sunfish are partially piscivorous and largemouth bass are partially insectivorous.

Aquatic ecology, as in terrestrial systems, is governed by the second law of thermodynamics. Approximately 10% of the energy in one trophic level is transferred to the next. Thus, the trophic status of the fish being produced in aquaculture determines the level of fish production (Table 18.1).

Trophic status of the cultured fish species also determines the variability associated with the effects of fertilization on fish production. The higher on the food chain that the fish production occurs, the less predictable the harvests become (Table 18.2).

Tilapia production in fertilized ponds is quite reliable due to the proximity of fish production to primary production. Larval walleye, which feed higher on the food chain and are thus farther removed from primary production, experience variable survival due to the probability of not intercepting the production of the zooplankton food source upon which they depend.

Table 18.1 Representative yields of fish at different trophic levels in ponds with equal fertility

Fish species	Trophic status	Typical fish yield (kg/Ha)
Tilapia	Phytoplanktivore	2500
Common carp	Benthic omnivore	800
Sunfish	Insectivore	250
Largemouth bass	Piscivore	100

Table 18.2 Intrinsic variability in aquaculture
yields of fish reared in fertilized ponds

Fish species	Coefficient of variation
Tilapia	20%
Sunfish	30%
Walleye	45%

It is common with walleye to have ponds with close to 100% survival adjacent to those with zero survival.

The highest production from fertilized ponds occurs in the poly-culture of Chinese carp, where several species with different low trophic level niches are reared together.

Organic fertilization

Organic fertilizer materials

Any material of direct or indirect plant origin, subject to microbial decomposition, qualifies as an organic fertilizer. Commonly used organic fertilizers are fresh or dried plant material, such as hay or straw, animal manures, seed residues, such as cottonseed cake or rice bran (Figure 18.1), and yeast.

There has been considerable research on the use of human waste from sewage treatment plants to produce fish. In nutrient-poor developing countries, the only available fertilizer material may be plant material that

Figure 18.1 Rice bran being applied to a tilapia nursery pond in Rwanda.

Figure 18.2 Where inorganic fertilizers and animal manures are unavailable, ponds may be fertilized with compost. This is a tilapia production pond in Rwanda.

is allowed to decompose in the pond (Figure 18.2). Sometimes a terrestrial crop is sowed in the dried pond bottom, later to be flooded and allowed to decompose. This is the common method of culturing crayfish, which feed directly on the decomposing organic matter. Crayfish and some fish species are grown in rice fields in order to take advantage of the straw left over from rice production (Figure 18.3).

Figure 18.3 Crayfish being raised in a harvested rice field in Louisiana.

Organic fertilizers are low in fertilizer nutrients, but high in carbon, and thus are most commonly used where heterotrophic production is desired over autotrophic production. For example, walleye ponds have traditionally been fertilized with organic materials, principally alfalfa hay, but sometimes supplemented with brewer's yeast or animal manure, because the decomposition of these materials has been thought to more reliably produce the zooplankton that these fish require than do the autotrophic pathway stimulated with large doses of soluble fertilizer nutrients. Organic fertilizers are also used when commercial inorganic fertilizers are unavailable or too expensive. Organic and inorganic fertilizers are often used together. The most common method for milkfish production is the use of poultry manure in combination with urea. Some organic fertilizers are too low in N for efficient or complete decomposition.

The most important characteristic of an organic fertilizer is its N content. The bacteria and other microbes that decompose organic matter are about 50% carbon (C) and 10% N, so N is required for microbial growth. If the organic matter being decomposed contains much N, microbes will grow well and some N will be mineralized into the environment. When organic fertilizer is low in N it must be immobilized from the environment in order for microbes to grow. This reduces soil N and N deficiency may limit organic matter decomposition. This point is illustrated by a simple experiment where the decomposition rates of plant materials of varying N contents are measured. The decomposition rate is measured in terms of oxygen consumption (Almazan and Boyd 1978; Table 13.1).

Because of the importance of N in organic matter decomposition, high N materials like leguminous plants, seed cakes, and animal manures are generally selected, when available, as organic fertilizers.

Organic fertilization rates

Many studies have shown that organic fertilization is as effective, or more effective, as the use of inorganic fertilizers in aquaculture (Swingle 1947; Hickling 1962; Collis and Smitherman 1978; Olah 1986; Schroeder et al. 1990; Opuszynski and Shireman 1993; Tice et al. 1996; Soderberg et al. 1997).

Pond management with organic fertilization is more labor intensive than with inorganic fertilizers due to the large quantities of material required for the same effect. Tice et al. (1996) found that organic fertilization was more expensive than inorganic fertilization at a state fish hatchery in Pennsylvania due to labor costs. In some areas, inorganic fertilizers are unavailable or too expensive, making the use of organic fertilizers more economical.

The mechanism by which organic fertilizers enhance pond productivity is microbial decomposition. Thus, heterotrophic production is an oxygen-consuming process and the major consideration in choosing organic fertilization rates is the prevention of oxygen depletion (Boyd 1990; Opuszynski and Shireman 1993; Lin et al. 1997; Das and Jana 2003). Tice et al. (1996) reported that walleye ponds receiving inorganic fertilizers had higher dissolved oxygen levels than those fertilized with organic matter at rates that resulted in equal fish production for both treatments. Lin et al. (1997) recommended that organic fertilizers be added daily to minimize the chances of oxygen depletion. Daily application is most practical in integrated farming operations, common in developing countries, where animal wastes are washed directly into fish ponds in daily cleaning operations. Hickling (1962) suggested that organic fertilizers be applied in piles, rather than spread over the entire pond bottom, to slow the rate of decomposition and hence spread the oxygen demand over a longer period of time. Because organic fertilizers vary in moisture content, usable recommendations on application rates with respect to oxygen depletion are based on dry weight. Other variables affecting organic fertilizers, not accounted for by using dry weights, are nutrient content and C:N ratios that are affected by the age of the material and resultant leaching of nutrients. Lin et al. (1997) reported that five batches of chicken manure from the same source in Thailand varied in moisture from 44% to 63%, in N content from 1.8% to 2.8%, and in P content from 2.6% to 3.5%. In an experiment where tilapia were grown in ponds fertilized with cattle manure, applications had to be limited to 80 kg/Ha dry matter per day in order to prevent oxygen depletion (Collis and Smitherman 1978). Boyd (1990) cautioned against the use of more than 50 kg/Ha dry matter per day for organic fertilization. Behrends et al. (1983) reported successful carp–tilapia polyculture when ponds were fertilized with 61 kg/Ha dry weight pig manure. The standard manuring rate in Europe is 100 kg/Ha dry weight per day (Olah 1986). Jhingran (1991) reported on the use of fresh cattle manure for carp production in India. Das and Jana (2003) estimated that the dry weight equivalent was approximately 106 kg/Ha dry matter per day for nursery ponds and 140–150 kg/Ha/day for grow-out ponds. Hickling (1962) reported that ponds in tropical areas are better able to assimilate large quantities of organic matter than those in more temperate locations. This may partly explain the conflicting reports on safe levels of organic fertilization.

Organic fertilization rates are generally given without regard to moisture content, so a few examples are provided here. In Java, where animal manures and inorganic fertilizers are not available, grass and leaves are applied to milkfish ponds at a rate of 1630 kg/Ha per application. Two or three such applications are required (Schuster 1952). Organic fertilization in the form of green manure is the principal method of culturing zooplankton-feeding game fish such as walleye. Typical fertilization rates range from 360 kg/Ha (Soderberg et al. 1997) to 800 kg/Ha (Summerfelt et al. 1993) of alfalfa applied as needed. Striped bass ponds are typically fertilized with cottonseed meal at rates of 225–560 kg/Ha with later applications of 56–170 kg/Ha (Geiger and Turner 1990).

Ponds used for carp polyculture in China are commonly fertilized with pig manure and grass. Typical application rates are 358 kg/Ha fresh pig manure and 1022 kg/Ha fresh aquatic grass per day (Zhang et al. 1987). In another description of Chinese polyculture, Tapiador et al. (1977) reported annual manure applications of 5625–10,125 kg/Ha in three applications. Carp nursery ponds in India were fertilized with 10,000 kg/Ha fresh cattle manure prior to stocking and 5000 kg/Ha 7 days post stocking (Jhingran 1991). Kapur and Lal (1986) reported that the maximum level of cattle manure fertilization was 10,000 kg/Ha.

Inorganic fertilization

Inorganic fertilizer materials

Any inorganic materials high in soluble amounts of nitrogen (N), phosphorus (P), and potassium (K), or combinations of such materials, qualify as inorganic fertilizers. Commercial fertilizers are classified according to their levels of N, as percent N; P, as percent P_2O_5; and K, as percent K_2O. A fertilizer labeled 20-10-5 contains 20% N, 10% P, as P_2O_5, and 5% K, as K_2O. The designation of P as P_2O_5 and K as K_2O is conventional rather than practical. The actual amounts of N, P, and K in 20-20-5, as percent N, P, and K, are 20, 8.7, and 2.5, respectively. Some researchers of fish pond fertilization have abandoned the archaic practice of recording P levels as P_2O_5 in favor of using percent P. Fertilizers containing N, P, and K are called complete fertilizers. Complete fertilizers are not often used in aquaculture because natural waters normally contain enough K to support aquatic primary productivity and *Cyanobacteria*, ubiquitous in aquatic systems, are capable of nitrogen fixation. A further source of N in pond waters, obviating the need for fertilizer N, is the mineralization from accumulated bottom soil organic matter. There is considerable evidence showing that fertilizer potassium is not needed in pond fertilizers (Hickling 1962; Dobbins and Boyd 1976; Das and Jana 2003). The lack of required N in fish pond fertilizers is less conclusive and will be discussed later in this chapter. Inorganic fertilizers are most commonly available in granular form, but liquid formulations are more effective for fish pond fertilization because of their increased nutrient solubility. Granular fertilizers can be dissolved in water prior to application to improve their effectiveness (Boyd 1990). Soderberg and Marcinko (1999) found that dissolved granular fertilizers were as effective as liquid formulations for the production of walleye. When dissolving granular fertilizers is impractical, they are best applied on a platform beneath the water surface (Figure 18.4).

Some commonly used fish pond fertilizers are listed in Table 18.3.

Figure 18.4 When dissolving granular fertilizers is impractical, they are best applied on a platform beneath the water surface. This is a milkfish pond in the Philippines.

Table 18.3 Some commonly used inorganic fish pond fertilizers

Fertilizer name	Form	Percentage			
		N	P_2O_5	P	K
Ammonium phosphate	Granular	11	46	20	0
Diammonium phosphate	Granular	18	46	20	0
Superphosphate	Granular	0	20	8.7	0
Triple superphosphate	Granular	0	46	20	0
Urea	Granular	45	0	0	0
Nitan plus[a]	Liquid	29	0	0	0
Ammonium polyphosphate	Liquid	10	34	14.8	0
Phosphoric acid	Liquid	0	54	23.5	0

[a] Agway, Westfield, Massachusetts.

The history of inorganic fertilization of fish ponds

Fish pond fertilization experiments in the United States began with H. S. Swingle in 1935 at Auburn University in Alabama. He fertilized farm ponds stocked with largemouth bass and bluegill with various amounts of complete fertilizers at approximately monthly intervals as indicated by increasing water clarity. In Swingle's early research he used only complete inorganic fertilizers and, thus, did not investigate the effects on fish production of individual nutrients. Fertilizer addition had marked effects on the production of largemouth bass and bluegills. Swingle's early research is summarized in Swingle (1947).

Hickling (1962), working at the Tropical Fish Culture Research Institute in Malacca, Malaya, conducted controlled, replicated experiments and showed that 22.4, 44.8, and 67.2 kg/Ha P_2O_5 (9.7, 19.5, and 29.2 kg/Ha P) resulted in increases in tilapia production of 298, 68, and 1% over the unfertilized control, respectively. Nitrogen and potassium had no effect on fish production. Hickling's work illustrates two important principles of inorganic fertilization of fish ponds. First, P is nearly always the limiting nutrient to primary production in water. Second, small doses of fertilizer result in large increases in fish production, while additional inputs increase fish production in ever-decreasing amounts. Thus, fertilization of fish ponds follows the law of diminishing returns (Figure 18.5) and the cost of fertilizer in relationship to the value of the fish will determine the most economic fertilization rate.

Swingle (1964) later compared complete fertilizers to those without N and found that fish production was unaffected by omission of N. A series of experiments conducted by C. E. Boyd and his students at Auburn University further demonstrated the importance of P in fish production and the small, if any, contribution of fertilizer N (Boyd and Sowles 1978; Murad and Boyd 1987). Hepher (1963) also showed the insignificant effect of fertilizer N on fish production.

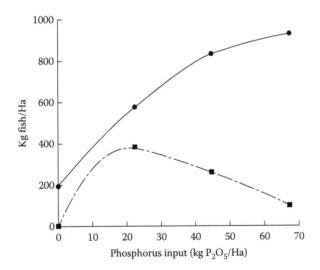

Figure 18.5 Tilapia response to fertilization with inorganic phosphorus showing diminishing returns from successive additions of fertilizer. Solid line is fish production. Dashed line is incremental increase in fish production per unit of fertilizer input. (Data from Hickling, C. F. 1962. *Fish Culture*. Faber and Faber, London, UK.)

Inorganic fertilization rates

The fertilization rate recommended by Swingle following the first 10 years of his research on pond fertilization (Swingle 1947) was 8–12 periodic additions of 112 kg/Ha of 6-8-4 and 11.2 kg of sodium nitrate (16-0-0). The purpose of the sodium nitrate addition was to bring the 6-8-4 fertilizer up to the 8-8-4 level that was available during the first years of his research. Later, when more concentrated complete fertilizers became available, the procedure was modified by Boyd and Snow (1975) to 8–12 periodic applications of 45 kg/Ha 20-20-5 (9 kg/Ha P_2O_5; 3.9 kg/Ha P). The total recommended phosphorus requirement for the entire February–November season was thus 72–100 kg P_2O_5 (31.3–43.5 kg/Ha P).

Hickling (1962) reported optimum tilapia production from the addition of 44.8 kg/Ha P_2O_5 (19.5 kg/Ha P) applied as a single dose at the beginning of a 6-month growing season. Dobbins and Boyd (1976) compared P fertilization rates by varying the P content of complete fertilizers while keeping the N and K levels constant. Ponds were stocked with small sunfish (*Lepomis* spp.) and fertilized 11 times during the growing period of January–November. This study (Table 18.4) showed that rather small amounts of P resulted in substantial increases in fish production and that higher fertilization rates did not result in statistically higher levels of fish production.

Sunfish are rather high on the food chain and thus experiments with these species typically result in high variability of fish response to fertilization. The numerical, but not significant, increases in P

Table 18.4 Response of sunfish to various levels of P fertilization

Year of experiment	Fertilizer treatment	Total amount added (kg/Ha)		Number of replicates per treatment	Sunfish production (kg/Ha)
		P_2O_5	P		
1976	20-5-5	24.8	10.8	4	80.3a
1976	20-10-5	49.5	21.5	4	197.6b
1976	20-15-5	74.3	32.3	4	210.4b
1976	20-20-5	99	43	4	270.0b
1976	20-20-0	99	43	4	220.2b
1977	20-10-5	49.5	21.5	6	322a
1977	20-20-5	99	43	6	360a

Source: 1976 data are from Dobbins, D. A. and C. E. Boyd. 1976. *Transactions of the American Fisheries Society* 105: 536–540; 1977 data are from Lichtkoppler, F. R., and C. E. Boyd. 1977. *Transactions of the American Fisheries Society* 106: 634–636.

Note: Means followed by the same letter in a given year are not significantly different ($P > 0.05$).

fertilization rates observed by Dobbins and Boyd (1976) were thought to be due to the small sample size of four replicates per treatment, so the experiment was repeated the following year (Lichtkoppler and Boyd 1977). In this trial, six replicates of two treatments, 20-20-5 and 20-10-5, were used. There was no significant difference in sunfish production between the two treatments, further demonstrating that previously recommended fertilization rates (Boyd and Snow 1975), while effective, were unnecessarily high. The final recommended annual fertilization rate was 49.5 kg/Ha P_2O_5 (21.5 kg/Ha P), or 4 kg/Ha P_2O_5 (1.7 kg/Ha P) per application.

Later experiments at Auburn University showed that liquid inorganic fertilizers were much more effective than granular forms. Metzger and Boyd (1980) found that 2.1 kg/Ha P_2O_5 (0.9 kg/Ha P) resulted in the same amount of sunfish production as 4 or 8 kg/Ha P_2O_5 (1.7–3.5 kg/Ha P) added in the granular form. Murad and Boyd (1987) found that phosphoric acid could be substituted for commercial liquid fertilizer, containing N, without affecting fish production. The superiority of liquid over granular fertilizers is explained by the increased solubility and subsequent availability of nutrients to plants. Granular fertilizer, while soluble, settles to the pond sediments, where much of the P is lost to the pond mud (Metzger and Boyd 1980). Boyd (1981) summarized the results of experiments comparing liquid and granular fertilizers (Table 18.5).

Lin et al. (1997) summarized 14 years of fertilization research on tilapia ponds at several locations worldwide by the Pond Dynamics/Aquaculture Collaborative Research Support Program (PD/A CRSP). Their recommendation was to fertilize with N at 2–4 kg/Ha/day and to fertilize with an N:P ratio of 4:1. Fertilization was most practical when the frequency of

Table 18.5 Summary of results conducted at Auburn University comparing liquid and granular fertilizers

	Rate (kg/Ha/application)			Sunfish production (kg/Ha)
Fertilizer	Total	P_2O_5	P	
Control	0	0	0	125a
Mixed (20-20-5)	45	9	3.9	228b
Triple superphosphate (0-46-0)	20	9	3.9	298b
Triple superphosphate	10	4.6	2	226b
Diammonium phosphate (18-46-0)	10	4.6	2	308b
Liquid fertilizer (15-25-0)	8	2	0.9	228b

Source: Data from Boyd, C. E. 1981. *Transactions of the American Fisheries Society* 110: 541–545.

Note: Means followed by the same letter are not significantly different (P > 0.05).

application was once every 2 weeks. Thus, the biweekly fertilization rate was 28–56 kg N and 7–14 kg P (16.1–32.2 kg P_2O_5)/Ha.

Wudtisin and Boyd (2005) used a regression procedure to determine that P fertilization in excess of 3 kg/Ha P_2O_5 (1.3 kg/Ha P) did not increase sunfish production. Regression of fish yields against fertilizer doses may be a more sensitive means of evaluating fertilizer requirements due to the great variability in treatment replicates.

Some researchers have recommended that fertilization rates be based on existing levels of dissolved N and P at the time of fertilizer application. Hepher (1963) fertilized carp ponds in Israel with a standard dose of 60 kg/Ha superphosphate (12 kg/Ha P_2O_5; 5.2 kg/Ha P) plus 60 kg/Ha ammonium sulfate (13 kg/Ha N) in order to maintain dissolved N and P concentrations of 0.5 and 2.0 mg/L, respectively. The optimum fertilization frequency was reported to be one standard dose every 2 weeks. The growing season in Israel was 6 months, so the annual fertilizer requirement was 144 kg/Ha P_2O_5 (62.6 kg/Ha P) and 156 kg/Ha N.

Lin et al. (1997) recommended that quantity and frequency of fertilization are best determined by measuring nutrient concentrations and making up the difference to some predetermined desired level or by conducting a bioassay of pond water to determine what nutrient levels result in the most prolific algal blooms.

Culver (1991) fertilized walleye ponds on a weekly basis with liquid inorganic fertilizers to maintain N and P concentrations of 30 and 600 µg/L, respectively. He hypothesized that small, unicellular algae that provide food for zooplankton could have a selective advantage over filamentous and colonial forms when nutrient levels were low. Culver (1991) found that the growth of filamentous *Cyanobacteria* was repressed and small flagellates increased when P concentrations were less than 30 µg/L. The addition of 30 µg/L equates to only 0.3 kg/Ha for a 1 m deep pond.

The procedures of formulating fertilization rates that involve frequent water quality analyses or bioassays are probably not practical for most aquaculture applications. Furthermore, fertilization rates based on dissolved nutrient concentrations ignore the influence of pond muds, which is substantial, especially with regard to P.

Comparison of inorganic fertilization rates

The foregoing analysis is by no means an exhaustive review of the literature on fish pond fertilization, but it describes many of the important contributions. It is evident that the rates described are variable and contradictory and that optimum fertilization rates are site-specific due to differences in climate, soil type, soil chemistry, water chemistry, and fish species selected for culture. A summary is provided (Table 18.6).

Table 18.6 Summary of fish pond inorganic phosphorus fertilization rate recommendations

Reference	Phosphorus requirement (kg/Ha) per application	
	P_2O_5	P
Swingle (1947)	9	3.9
Boyd and Snow (1975)	9	3.9
Hickling (1962)	3.7	1.6
Hepher (1963)	12	5.2
Dobbins and Boyd (1976)	4	1.7
Lichtkopler and Boyd (1977)	4	1.7
Metzger and Boyd (1980) (Liquid)	2.1	0.9
Culver (1991)	1.2	0.5
Lin et al. (1997)	14–28	6.1–12.2
Wudtisin and Boyd (2005)	3	1.3

Note: Hickling's (1962) 6-month rate was divided by 12 for comparison to other studies that reported results based on 2-week fertilization intervals. Culver's (1991) 1-week rate was multiplied by two to compare it to a 2-week interval.

Importance of inorganic nitrogen fertilization

Many studies have failed to demonstrate a benefit to aquatic productivity from fertilizer nitrogen (Hickling 1962; Hepher 1963; Swingle 1964; Boyd and Sowles 1978; Murad and Boyd 1987). Nonetheless, recent fertilizer recommendations have included N (Culver 1991; Lin et al. 1997; Boyd et al. 2008). It is generally concluded that nitrogen fixation by *Cyanobacteria* is sufficient to satisfy the needs for aquatic production and, thus, N is not a limiting factor. Unnecessary N fertilization can be harmful. Ammonium- and urea-based fertilizers are acid forming due to the release of hydrogen ions during nitrification. Furthermore, ammonium and urea release ammonia to the water, which becomes toxic to fish when plankton respiration raises pH levels of pond waters. The adverse effects of elevated NH_3 exposure on fish health and growth in fertile static water ponds have been recorded by Soderberg et al. (1983) and Soderberg (1985).

Hepher (1963) added fertilizer N to elevate dissolved levels of N to 2.0 mg/L, the concentration previously found in ponds with optimum primary productivity, but he did not experiment with P fertilization alone.

Lin et al. (1997) point out that the studies showing little or no benefit from fertilizer N were in systems where production was rather low or the systems were stocked with nonmicrophagus filter-feeding species. Natural nitrogen fixation may not be sufficient to support the hyper-eutrophic tilapia ponds described in the PD/A CRSP reports (Lin et al. 1997). Nile tilapia is a microphagus filter feeder.

Culver (1991) recommended N fertilization not to enhance primary productivity, but to manipulate the algal species composition. He hypothesized that low levels of N could favor *Cyanobacteria* over the more palatable green algae and that relatively high levels of N fertilization with low levels of P could produce an algal community more favorable to the production of zooplankton and, hence, walleye, which feed on them.

Boyd et al. (2008) studied N fertilization in Alabama sunfish ponds that had recently been renovated by dredging. When pond bottom sediments containing the residues of many years of annual fertilization were removed, N fertilization had a significant effect on fish production. The resulting fertilization recommendation for newly renovated ponds or those without a history of fertilization was 3 kg/Ha P_2O_5 (1.3 kg/Ha P) and 6 kg/Ha N per application. Ponds with a fertilization history of 5 years or more did not require N fertilization.

Use of combinations of organic and inorganic fertilizers

Boyd (1990) summarized the literature on pond fertilization and concluded that a combination of organic and inorganic fertilizers is more effective than either treatment alone. Geiger et al. (1985) showed that a combination of organic and inorganic fertilizers was much better than inorganic fertilizer alone for producing zooplankton forage for larval striped bass. The traditional method of fertilization of milkfish ponds in the Philippines is with chicken manure and urea. This procedure has been found to be the best method for culturing the rich benthic community upon which the fish graze. Ponds used for the production of zooplankton-feeding game fish are often fertilized with combinations of organic and inorganic materials (Fox et al. 1992; Anderson 1993; Myers et al. 1996).

Reasons for the benefit of mixed fertilizer strategies include the provision of a more complex and thus more predictable food production mechanism, but the largest benefit is probably from the addition of N to N-deficient organic material that accelerates its decomposition. Urea is often added to ponds fertilized with organic fertilizers in order to supply sufficient N for decomposition. In one experiment with tilapia ponds in Thailand, 200 kg/Ha chicken manure supplemented with 22.4 kg/Ha urea produced greater yields of tilapia than 1000 kg/Ha of chicken manure alone (Diana et al. 1991). Ganguly et al. (1999) clearly demonstrated that a combination of animal manures and inorganic fertilizer was better than manures alone. They attributed this result to improved N:P ratios provided by the inorganic fertilizer. Dinesh et al. (1986) recommended that

2000 kg/Ha of poultry manure supplemented with 100 kg/Ha urea was a safe and economical fertilization rate for carp ponds.

Advantages and disadvantages of organic and inorganic fertilizers

Inorganic fertilizers are nutrient rich and soluble, providing large doses of fertilizer nutrients for immediate uptake by primary producers. Organic fertilizers are nutrient poor and must decompose before effectively stimulating the mainly heterotrophic pathways that they support. The decomposition rate of the organic material is variable and unpredictable. The most important factor determining the decomposition rate is the N content. Thus, inorganic nitrogen is often added to organic fertilizers to improve their effectiveness.

In developed countries, organic fertilization is more expensive than inorganic fertilization because of the high labor costs of applying the large quantities of material required. Where labor costs are lower, organic fertilization is generally less expensive than purchasing manufactured inorganic fertilizers. In some locations, inorganic fertilizers are unavailable, making organic fertilizers the only option.

There is a risk of oxygen depletion in hyper-eutrophic inorganically fertilized ponds due to plankton respiration, but dissolved oxygen problems are much more prevalent in ponds fertilized with organic fertilizers due to the oxygen requirement for decomposition. Care must be taken when adding organic matter to ponds. Relatively small, daily additions are usually recommended.

Most inorganic fertilizers containing N hydrolyze to ammonia, which can lead to fish toxicity at the high afternoon pH levels typical of fertile fish ponds.

Inorganic fertilizers have a drastic and immediate effect on primary production from phytoplankton. Organic fertilizers require bacteria and other microbes for decomposition and thus offer a wider diversity of fish foods, particularly zooplankton. Combinations of organic and inorganic fertilizers not only improve the response of the organic matter but provide both inorganic and organic components of fertilization.

Organic fertilizers have potential public health and aesthetic concerns. Excessive manuring can create an environment favorable to pathogenic bacteria. Several studies have related organic fertilization to the proliferation of helminthes that require fish or other pond organisms as intermediate hosts of human diseases (Larrson 1994; Santos 1994; Polprasert 1996).

Boyd (2003) cautioned against the use of animal manures because of their oxygen demand, the deterioration of pond bottom soil conditions, and the possibility that they may contain heavy metals or antibiotics. There is a growing concern over the use of animal manures for the production of food fish due to aesthetic and sanitary concerns (Tucker et al. 2008).

SAMPLE PROBLEMS

1. Consider bacteria composed of 50% C and 10% N with a carbon assimilation efficiency of 7%. Carbon assimilation efficiency means how much decomposed carbon is assimilated into bacterial cells. The rest is given off as CO_2 in respiration. If these bacteria decompose 100 kg of straw that contains 40% C and 0.1% N, how much nitrogen will be mineralized into or immobilized from the pond mud?

2. If 1 kg of inorganic nitrogen is added to the straw in Problem 1, how much nitrogen will be mineralized or immobilized? How much bacterial biomass will be produced?

3. A manure contains 45% C and 1.1% N. How much nitrogen will be mineralized or immobilized when 1 tonne of this material is decomposed by bacteria with an 8% carbon assimilation efficiency?

4. How could the addition of fertilizer P to a fish pond increase the amount of nitrogen available for plant growth?

5. You are instructed to fertilize a 10.5 Ha pond with super phosphate at a rate of 50 kg/Ha P_2O_5. How much do you need?

6. Devise a way to fertilize a pond with 25 kg/Ha N and 45 kg/Ha P_2O_5 using liquid materials.

7. You are provided with the following experimental data:

Treatment (kg/Ha P_2O_5)	Yield (kg/Ha)
Control	97
22.4	317
44.8	418

If fish are worth $1.00/kg and ammonium phosphate costs $1.00/kg, estimate the optimum fertilization rate.

8. How could you make a fertilizer with the nutrient composition of 20-20-0?

9. Monoammonium phosphate is 11-48-0. How could you make this into a fertilizer that contains 20% N?

10. Ponds used in the production of juvenile walleye are fertilized with 30 mg/L P and 600 mg/L N at approximately 2-week intervals. How much phosphoric acid and Nitan Plus are required to treat a 0.5 Ha, 1 m deep walleye pond?

References

Almazan, G. and C. E. Boyd. 1978. Effects of nitrogen levels on rates of oxygen consumption during decay of aquatic plants. *Aquatic Botany* 5: 119–126.

Anderson, R. O. 1993. Apparent problems and potential solutions for production of fingerling striped bass, *Morone saxatilis*. Pages 119–150 *in* R. O. Anderson and D. Tave, editors. *Strategies and Tactics for Management of Fertilized Hatchery Ponds*. Food Products Press, Binghamton, New York.

Behrends, L. L., J. B. Kingsley, J. J. Maddox, and E. L. Waddell Jr. 1983. Fish production and community metabolism in an organically fertilized fish pond. *Journal of the World Mariculture Society* 14: 510–522.

Boyd, C. A., P. Penseng, and C. E. Boyd. 2008. New nitrogen fertilization recommendations for bluegill ponds in the southeastern United States. *North American Journal of Aquaculture* 70: 308–313.

Boyd, C. E. 1981. Comparison of five fertilizer programs for fish ponds. *Transactions of the American Fisheries Society* 110: 541–545.

Boyd, C. E. 1990. *Water Quality in Ponds for Aquaculture*. Alabama Agricultural Experiment Station, Auburn University, Auburn, Alabama.

Boyd, C. E. 2003. Bottom soil and water quality management in shrimp ponds. Pages 11–33 *in* B. B. Jana and C. D. Webster, editors. *Sustainable Aquaculture: Global Perspectives*. Food Products Press, Binghamton, New York.

Boyd, C. E. and J. R. Snow. 1975. *Fertilizing Farm Fish Ponds*. Leaflet 88, Alabama Agricultural Experiment Station, Auburn University, Auburn, Alabama.

Boyd, C. E. and J. W. Sowles. 1978. Nitrogen fertilization of ponds. *Transactions of the American Fisheries Society* 107: 737–741.

Collis, W. J. and R. O. Smitherman. 1978. Production of tilapia hybrids with cattle manure or a commercial diet. Pages 43–54 *in* R. O. Smitherman, W. L. Shelton, and J. H. Grover, editors. *Culture of Exotic Fishes Symposium Proceedings*. Fish Culture Section, American Fisheries Society, Auburn, Alabama.

Culver, D. A. 1991. Effects of the N:P ratio in fertilizer for fish hatchery ponds. *Verrhandlungen Internationale Vereinigung fur Theoretische und Angewandte Limnologie* 24: 1503–1507.

Das, S. K. and B. B. Jana. 2003. Pond fertilization regimen: State-of-the-art. *Journal of Applied Aquaculture* 13: 35–66.

Diana, J. S., C. K. Lin, and P. J. Schneeberger. 1991. Relationship among nutrient inputs, water nutrient concentrations, primary productivity and yield of *Oreochromis niloticus* in ponds. *Aquaculture* 92: 323–341.

Dinesh, K. R., T. J. Varghese, and M. C. Nandeesha. 1986. Effects of a combination of poultry manure and varying doses of urea on the growth and survival of cultured carps. Pages 565–568 *in* J. L. MacLean, L. B. Dizon, and L. V. Hosillos, editors. *Proceedings of the First Asian Fisheries Forum*. Manila, Philippines.

Dobbins, D. A. and C. E. Boyd. 1976. Phosphorus and potassium fertilization of sunfish ponds. *Transactions of the American Fisheries Society* 105: 536–540.

Fox, M. G., D. D. Flowers, and C. Waters. 1992. The effect of supplementary inorganic fertilization on juvenile walleye (*Stizostedion vitreum*) reared in organically fertilized ponds. *Aquaculture* 106: 27–40.

Ganguly, S., J. Chatterjee, and B. B. Jana. 1999. Biogeochemical cycling bacterial activity in response to lime and fertilizer applications in pond systems. *Aquaculture International* 7: 413–432.

Geiger, J. C., C. J. Turner, K. Fitzmayer, and W. C. Nichols. 1985. Feeding habits of larval and fingerling striped bass and zooplankton dynamics in fertilized rearing ponds. *Progressive Fish-Culturist* 47: 213–223.

Geiger, J. G. and C. J. Turner. 1990. Pond fertilization and zooplankton management techniques for production of fingerling striped bass and hybrid striped bass. Pages 79–98 *in* R. M. Harrell, J. H. Kerby, and R. V. Minton, editors. *Culture and Propagation of Striped Bass and Its Hybrids.* Striped bass committee, Southern Division, American Fisheries Society, Bethesda, Maryland.

Hepher, B. 1963. Ten years of research in fish pond fertilization in Israel. II. Fertilizer dose and frequency of fertilization. *Bamidgeh* 15: 78–92.

Hickling, C. F. 1962. *Fish Culture*. Faber and Faber, London, United Kingdom.

Jhingran, V. G. 1991. *Fish and Fisheries in India*. Hindustan Publishing Corporation, Delhi, India.

Kapur, K. and K. K. Lal. 1986. The chemical quality of waste treated waters and its relation with patterns of zooplankton populations. Pages 129–132 *in* J. L. MacLean, L. B. Dizon, and L. V. Hosillos, editors. *Proceedings of the First Asian Fisheries Forum*. Asian Fisheries Society, Manila, Philippines.

Larrson, B. 1994. *The Overviews on Environment and Aquaculture in the Tropics and Subtropics*. ALCOM Field Document No 27, FAO, Rome, Italy.

Lichtkoppler, F. R. and C. E. Boyd. 1977. Phosphorus fertilization of sunfish ponds. *Transactions of the American Fisheries Society* 106: 634–636.

Lin, C. K., D. R. Teichert-Coddington, B. W. Green, and K. L. Veverica. 1997. Fertilization regimes. Pages 73–107 *in* H. S. Egna and C. E. Boyd, editors. *Dynamics of Pond Aquaculture*. CRC Press, Boca Raton, Florida.

Metzger, R. J. and C. E. Boyd. 1980. Liquid ammonium polyphosphate as a fish pond fertilizer. *Transactions of the American Fisheries Society* 109: 563–570.

Murad, A.,and C. E. Boyd. 1987. Experiments on fertilization of sport-fish ponds. *Progressive Fish-Culturist* 49: 100–107.

Myers, J. J., R. W. Soderberg, J. M. Kirby, and M. T. Marcinko. 1996. Production of walleye (*Stizostedion vitreum*) in earthen ponds fertilized with organic and inorganic fertilizers and stocked at three rates. *Journal of Applied Aquaculture* 6: 11–19.

Olah, J. 1986. Carp production in manured ponds. Pages 295–303 *in* R. Billard and J. Marcel, editors. *Aquaculture of Cyprinids*. Institut National de la Recherche Agronomique, Paris, France.

Opuszynski, K. K. and J. V. Shireman. 1993. Strategies and tactics for larval culture of commercially important carp. Pages 189–220 *in* R. O. Anderson and D. Tave, editors. *Strategies and Tactics for Management of Fertilized Hatchery Ponds*. Food Products Press, Binghamton, New York.

Polprasert, C. 1996. *Organic Waste Recycling*. John Wiley & Sons, Chichester, England.

Santos, C. A. L. 1994. Prevention and control of food borne trematode infections in cultured fish. *FAO Aquaculture Newsletter* 8: 11–15.

Schroeder, G. L., G. Wohlfarth, A. Alkon, A. Halevy, and H. Krueger. 1990. The dominance of algal-based food webs in fish ponds receiving chemical fertilizers plus organic manures. *Aquaculture* 86: 219–229.

Schuster, W. H. 1952. *Fish Culture in Brackish Water Ponds of Java*. Indo-Pacific Fisheries Council, Special Publication 1, Bangkok, Thailand.

Soderberg, R. W. 1985. Histopathology of rainbow trout, *Salmo gairdneri* (Richardson), exposed to diurnally fluctuating un-ionized ammonia levels in static-water ponds. *Journal of Fish Diseases* 8: 57–64.

Soderberg, R. W., J. B. Flynn, and H. R. Schmittou. 1983. Effects of ammonia on the growth and survival of rainbow trout in intensive static-water culture. *Transactions of the American Fisheries Society* 112: 448–451.

Soderberg, R. W., J. M. Kirby, D. Lunger, and M. T. Marcinko. 1997. Comparison of organic and inorganic fertilizers for the pond production of walleye (*Stizostedion vitreum*). *Journal of Applied Aquaculture* 7: 23–29.

Soderberg, R. W. and M. T. Marcinko. 1999. Substitution of granular for liquid fertilizers for the pond production of walleye, *Stizostedion vitreum*, in earthen ponds. *Journal of Applied Aquaculture* 9: 33–40.

Summerfelt, R. C., C. P. Clouse, and L. M. Harding. 1993. Pond production of fingerling walleye, *Stizostedion vitreum*, in the northern great plains. Pages 33–58 *in* R. O. Anderson and D. Tave, editors. *Strategies and Tactics for Management of Fertilized Hatchery Ponds*. Food Products Press, Binghamton, New York.

Swingle, H. S. 1947. *Experiments on Pond Fertilization*. Bulletin 264, Alabama Agricultural Experiment Station, Alabama Polytechnical Institute, Auburn, Alabama.

Swingle, H. S. 1964. Pond fertilization, bluegill—Bass with fertilization. Pages 62–64 *in Fisheries Research Annual Report*. Alabama Agricultural Experiment Station, Auburn University, Auburn, Alabama.

Tapiador, D. D., H. F. Henderson, M. N. Delmwndo, and H. Tsutsui. 1977. Freshwater fisheries and aquaculture in China. FAO Fisheries Technical Papers 168, Rome, Italy.

Tice, B. J., R. W. Soderberg, J. M. Kirby, and M. T. Marcinko. 1996. Growth and survival of walleye (*Stizostedion vitreum*) reared at two stocking rates in ponds fertilized with organic and inorganic materials. *Progressive Fish-Culturist* 58: 135–139.

Tucker, C. S., J. A. Hargreaves, and C. E. Boyd. 2008. Better management practices for freshwater pond aquaculture. Pages 151–226 *in* C. S. Tucker and J. A. Hargreaves, editors. *Environmental Best Management Practices for Aquaculture*. Wiley-Blackwell, Ames, Iowa.

Wudtisin, W. and C. E. Boyd. 2005. Determination of the phosphorus fertilization rate for bluegill ponds using regression analysis. *Aquaculture Research* 36: 593–599.

Zhang, F. L., Y. Zhu, and X. Y. Zhow. 1987. Studies on the ecological effects of varying the size of fish ponds loaded with manures and feeds. *Aquaculture* 60: 107–116.

chapter nineteen

Use of artificial diets in
static water aquaculture

The rapid increase in aquaculture production and the elevation of tilapia, striped catfish, and marine shrimp from local to global commodities has occurred mainly due to the widespread availability and utilization of commercial dry fish diets. Total world fish production was 158 million tonnes in 2012, 66.6 million tonnes of which were food fish produced in aquaculture (FAO 2014a,b). This total includes finfish, crustaceans, mollusks, and other edible animals. The rapidly increasing use of commercial diets has led to dramatic increases in food fish production for aquaculture over the past decade. Farmed fish contributed 13.4%, 25.7%, and 42.2% of the total world fish production in 1990, 2000, and 2012, respectively (FAO 2014a,b). Of the total aquaculture production in 2012, approximately 46 million tonnes resulted from the use of artificial diets. Non-fed species included filter-feeding carp (7.1 million tonnes) and bivalves (13.4 million tonnes). The share of production dependent on artificial diets increased from 66.5% in 2010 to 69.2% in 2012, reflecting an increasing trend in the use of commercial fish diets (FAO 2014a,b). Aquaculture products that have evolved from domestic to global commodities in recent years include Atlantic salmon, tilapia, striped catfish (swai), and marine shrimp. With the exception of Atlantic salmon, all of these are cultured primarily in static water earthen ponds.

Artificial fish diets

The first diets used to increase production over that possible with the manipulation of natural foods were low in cost, low in protein and micronutrients, and high in energy. They did not satisfy all the nutritional requirements of the fish, but increased production by sparing the protein present in natural foods. Such supplemental diets included rice bran and other grain products. As fish production intensified, more complete diets were required to supply the nutritional deficiencies of the natural/supplemental diet combination. Modern fish diets are complete in their nutritional composition and can be used in tanks and cages where natural foods are not present.

Fish have high protein requirements and fish diets must contain the 10 essential amino acids (arginine, histidine, isoleucine, leucine, lysine, methionine, phenylalanine, threonine, tryptophan, and valine) that fish are unable to synthesize from other products (Lim and Webster 2006). The amino acid profiles of plant proteins are deficient in essential amino acids and even tilapia, which are generally considered to be vegetarian, require some animal protein for adequate growth and health. Fish meal is used in most fish diets because of its amino acid profile, but one of the most important issues in the sustainability of rapidly increasing aquaculture production is the availability of fish meal and the effect of the demand for fish meal on marine fishery resources. An important component of fish nutrition research has been efforts to reduce the fish meal requirements of fish diets. Half the fish meal requirement in tilapia diets was successfully replaced with poultry by-product meal (Lim and Webster 2006) and spray-dried blood meal (Lee and Bai 1997). Shiau et al. (1987) and Tacon et al. (1983) showed that soybean meal enriched with synthetic methionine could reduce some of the fish meal requirements of tilapia diets.

Warmwater fish generally require less protein than coldwater fish because protein-sparing fats and carbohydrates are more digestible at warmer temperatures and coldwater fish are at a higher trophic status than warmwater fish. Juvenile fish require more protein than adult fish. For example, striped catfish (*Pangasianodon hypophthalmus*) from 5–50 g require 34%–36% protein, 50–100 g fish require 32%–34%, 300–500 g fish require 24%–26%, and fish over 500 g require 2%–26% protein (Glencross et al. 2011). Tilapia (*Oreochromis niloticus*) under 1 g require 40% dietary protein (Siddiqui et al. 1988), while 9 g fish require 25% (Wang et al. 1985). Typical grow-out diets for tilapia contain 32% protein (Li et al. 2006). In contrast, modern trout diets contain 40%–42% protein.

Most of the energy in fish diets is supplied by lipids. Tilapia can tolerate dietary lipids at levels as high as 12% (Jauncey 2000). Modern salmonid diets contain 16%–20% fat because of its high energy content, lower cost than protein, and ability to spare protein. Fish require fatty acids of the omega 3 and 6 (n-3 and n-6) families. Corn and soybean oils are good sources of n-6 fatty acids; omega 3 fatty acids are added to fish diets in the form of fish oil.

Carbohydrates are the least expensive energy sources included in fish diets and fish, especially those of lower trophic status such as tilapia and carp, digest carbohydrates relatively well. Fish do not require dietary carbohydrates, but they are always added to fish diets because of their low cost, (albeit limited) digestibility, their ability to spare protein, and their function as binding agents in feed pellet manufacture.

Complete fish diets require the addition of vitamins and minerals. When fish are fed otherwise complete diets lacking vitamins, nutritional deficiencies are not common at low fish densities due to the presence of natural feeds (Lovell 1989). Vitamin C deficiencies occur in channel catfish ponds at densities above 3000 kg/Ha. When Prather and Lovell (1971) added a vitamin premix designed for poultry to channel catfish diets, fish weight increased by 19% over control diets without the vitamin addition. All commercial fish diets now contain a vitamin premix.

Fish diet components are blended, ground, and formed in stable pellets in one of two ways. Steam pelleting is the simpler process and produces a sinking pellet. Extrusion processing produces a floating pellet. In the manufacturing process pressure is applied and suddenly reduced so that moisture in the feed mixture vaporizes, forming air pockets and expansion of the pellets. Steam pelleting is most common in developing counties because the necessary machinery is simpler and less expensive than the cooker-extruder used in the manufacture of floating feeds. Floating pellets are desirable because they allow for better observation of feeding behavior. Most finfish readily accept floating pellets, but shrimp require a sinking pellet.

The process of calculating the feeding rate based on anticipated growth and food conversion used in flowing water applications (see Chapter 2) is rarely used in static water aquaculture. This is because feeding behavior is influenced by temperature, dissolved oxygen (DO), ammonia exposure, and other environmental factors that change dielly and seasonally in static water ponds. Furthermore, fish biomass in ponds is usually not accurately known. Thus, feeding fish in ponds is more of a skill than a science. The skilled fish feeder knows from experience how to assess fish appetite, which is highly variable. Fish growth is usually best when fish are fed several times per day to satiation, and satiation is reached in 20–30 minutes following initiation of feeding (Tucker et al. 2008). Feeding rates range from 1%–4% of body weight per day depending upon fish size and water temperature. Small fish require a higher feeding rate than larger fish. Maximum feeding rates are determined by deterioration in water quality rather than the appetites of the fish.

Effects of feed addition on water quality

Most of the published information on feed-based static water aquaculture concerns the American channel catfish industry. I will therefore use the development and intensification of that industry as a general example. When prepared pelleted diets are added to fish ponds the result is eutrophication. Because the amount of added feed is unlimited, water quality degradation is much more important in fed than fertilized ponds. The degree of eutrophication caused by artificial feeding is proportional to the amount of feed provided. The proliferation of plankton resulting from the addition of fish feed causes increasingly wide fluctuations in DO due

to plankton photosynthesis and respiration. In these cases oxygen depletion can become so severe that it may result in fish mortality.

A further consequence of wildly fluctuating DO levels is daily
elevations of pH, which results in short term exposure of fish to
potentially lethal levels of NH_3. The chemistry and toxicity of ammonia
are described in Chapter 9.

Furthermore, as culture intensity increases, nitrite (NO_2^-) accumulates in pond waters. Nitrite is the intermediate product of nitrification
and can accumulate to toxic levels in aquaculture systems. The mode of
toxicity of NO_2^- is conversion of hemoglobin to methemoglobin, which
cannot carry oxygen. The resulting disease is called methemoglobinemia.
The blood of affected fish is brown in color. Losordo (1997) recommended
that NO_2^- levels be kept below 5 mg/L, but channel catfish exposed to
this level for 24 hours showed a conversion of 77% of their hemoglobin
to methemoglobin (Tomasso et al. 1979). It is impossible to assign a maximum tolerable concentration for NO_2^- to fish because its toxicity is highly
variable depending upon pH, chloride, and calcium levels. Nitrite toxicity
in ponds can be controlled by chloride addition (Perrone and Meade
1977; Tomasso et al. 1979) and fish in brackish or seawater are resistant
to methemoglobinemia (Almendras 1987). Recommendations on effective NO_2^-:Cl^- ratios range from 3:1 (Schwedler et al. 1985) to 20:1 (Losordo
1997). Because Ca^{2+} also protects fish from NO_2^- toxicity (Wedemeyer and
Yasutake 1978), $CaCl_2$ is more effective than NaCl for this purpose.

Generally, feeding rates of up to 30 kg/Ha per day can be maintained without danger of fish mortality caused by nighttime oxygen
depletion. This level of feeding results in a maximum fish crop of around
2000–3000 kg/Ha, depending upon the trophic status of the aquaculture species, which is generally considered to be unacceptably low for
economical production. Aeration is the most common and most effective
method of preventing fish kills caused by excessive plankton respiration
resulting from feeding rates in excess of 30 kg/Ha.

Aeration of static water fish ponds

When the channel catfish industry had developed to the level of intensity that required occasional emergency aeration, the aeration was accomplished with tractor-driven paddlewheel units that could be moved from
pond to pond as needed (Figures 19.1 and 19.2). In order to effectively
deploy the available paddlewheel units, a method for predicting which
ponds would require emergency aeration was required so that units could
be deployed if early morning DO levels were likely to fall below 2–3 mg/L.

Boyd et al. (1979a) presented a mathematical procedure in which
the individual components of the nighttime DO budget were quantified
and then combined to produce a total oxygen budget. Diffusion is the

Figure 19.1 Tractor-driven paddlewheel aerator used for emergency aeration in semi-intensive static water aquaculture.

Figure 19.2 Small tractor-driven aerators used to aerate sport fish ponds.

loss or gain of DO to or from the atmosphere, resulting from the oxygen pressure gradient between the water and air. Data from Schroeder (1975) were used that showed 12-hour changes in DO when dusk DO levels were 50%–250% saturation. Subjecting these data to regression analysis results in the following expression:

$$DF = -0.024\,DO + 2.677 \qquad (19.1)$$

where $DF = DO$ change in mg/L and $DO = DO$ at dusk in percent saturation.

Benthic respiration rates are difficult to measure and vary widely so Boyd et al. (1979a) suggested that 61 mg DO/m^2 hour be used as this was the best available average value for the benthic respiration rate (Mezainis 1977).

The following expression (Andrews and Matsuda 1975) was used in the model for fish respiration:

$$\log O_2 = -0.999 - 0.000957\,W + 0.0000006\,W^2 + 0.0327\,T$$
$$- 0.0000087\,T^2 + 0.0000003\,WT \qquad (19.2)$$

where O_2 = oxygen consumption rate in mg DO/g fish hour, W = average fish weight in g, and T = water temperature in degrees C.

The biological oxygen demand (BOD) is the oxygen demand of plankton and bacteria in the water column. Romaire and Boyd (1978) give the following equation based on Secchi disc visibility:

$$BOD = -1.133 + 0.00381\,S + 0.0000145\,S^2 + 0.0812\,T - 0.000749\,T^2$$
$$- 0.000349\,ST \qquad (19.3)$$

where BOD = mg DO/L hour, S = Secchi disc visibility in cm, and T = water temperature in degrees C. The total combined oxygen budget was, thus,

$$DO_{dawn} = DO_{dusk} \pm DF - DO_{fish} - DO_{mud} - BOD \qquad (19.4)$$

A series of tables was published by the Auburn University Agriculture Experiment Station that allowed the fish farmer to input dusk DO, estimated fish biomass and average fish weight, and Secchi disk visibility, and read off the estimated dawn DO level so the calculations would not have to be performed for each estimate.

The pond oxygen budget method is not in widespread use at the present time, but it provides insight for the understanding of the nighttime respiration process and the relative contribution of each component.

In the preceding example, a 1Ha pond, 1 m deep contains 4000 kg of 400 g fish. The dusk DO is 14 mg/L, S = 20 cm, and T = 28°C.

The solution to the problem follows:

DO saturation at 28°C is 7.36 mg/L (Table 6.1) so 14 mg/L = 190% saturation
$DF = -1.88$ mg/L
$DO_{fish} = 1.24$ mg/L
$DO_{mud} = 61$ mg/L hour $\times$ 12 hour/1000 L = 0.73 mg/L

BOD = 0.437 mg/L hour × 12 hour = 5.24 mg/L

DO_{dawn} = 14 − 1.88 − 1.24 − 0.73 − 5.24 = 4.91 mg/L. Aeration will not be required.

Boyd et al. (1979a) presented a simpler and equally effective method of predicting early morning DO levels and the need to deploy emergency aeration, taking at least two DO measurements at different times, after dark, and project the DO at dawn using a straight line through the two measured points (Figure 19.3).

The method of managing channel catfish production, primarily in Mississippi and Alabama, until approximately the end of the 1970s, involved the monitoring of DO and deployment of emergency aeration as necessary. Typically, all ponds were checked every afternoon and every morning. If the morning DO was 5 mg/L or greater and the afternoon DO was higher than it had been the evening before, no nighttime surveillance was necessary. If the evening DO was falling and dawn DOs were usually low, DO was checked at intervals throughout the night and the projection method (Figure 19.3) was used to determine whether emergency aeration would be required. This level of management is now referred to as semi-intensive. Boyd et al. (1979b) studied semi-intensive catfish culture in actual production ponds. Replicate ponds were stocked with 5000, 10,000, and 20,000 fish per Ha so that feeding rates reached 34, 56,

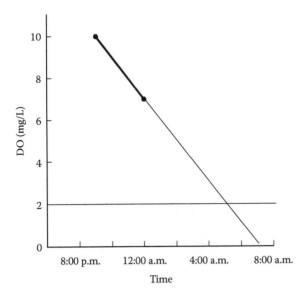

Figure 19.3 Projection method of predicting early morning DO levels. Measurements of DO were 10 mg/L at 8:00 p.m. and 7 mg/L at 12:00 a.m. Aeration will be required at approximately 5:00 a.m. when DO falls to 2 mg/L.

and 78 kg/Ha day. In the low treatment there were no DO problems, the fish survival rate was 99%, fish production was 3000 kg/Ha, and profit was $1136/Ha. In the medium treatment, one pond in six experienced DO problems and some fish died despite the use of emergency aeration. Survival still averaged 93%. Fish production was 4100 kg/Ha and profit was $1303/Ha. In the high treatment, some fish died in three of the six ponds and survival averaged 83%. Fish production was 4900 kg/Ha, but profit was only $671/Ha due to fish mortality and reduced fish size. A maximum feeding rate of 56 kg/Ha was the most economical and, with careful use of aeration, the risk of fish kills was acceptable. At 78 kg/Ha, emergency aeration was required every night and suitable conditions for fish health and growth were difficult to maintain. The typical annual production limit for American catfish production using semi-intensive methods with portable emergency aeration equipment is 3000–4500 kg/Ha (Boyd 1998).

By the late 1970s, American catfish farmers had increased stocking and feeding rates to the point that aeration was required most nights in most ponds (Torrens 2005) and the use of permanently installed electric aerators (Figure 19.4) became common (Busch et al. 1984). Intensive culture of channel catfish employs aeration with dedicated electric paddlewheels at an aeration rate of 3.5–4.7 kW/Ha (USDA 1997).

Cole and Boyd (1986) studied the production limits of the intensive method of channel catfish production by stocking fish at six rates from 1200 to 34,600/Ha so that maximum daily feeding rates reached 224 kg/

Figure 19.4 Permanently placed electric paddlewheel aerator used for intensive static water aquaculture shown in drained pond. (Photograph courtesy of Lauren Jescovitch.)

Ha day in the high treatment. The study was conducted in small experimental ponds with dedicated aeration at a rate of 6.1 kW/Ha. Aerators were deployed when DO concentrations were predicted to fall below 2 mg/L. At feeding rates of 168 and 224 kg/Ha/day DO was often depleted, or nearly depleted by dawn, even though the aerators were running. At feeding rates over 112 kg/Ha/day aerators were operated almost continuously between midnight and dawn in August and September. Aeration at a rate of 6.1 kW/Ha did not provide sufficient DO to ensure fish health at feeding rates above 112 kg/Ha day and fish production decreased as a result. Food conversion rates increased due to the inability of the fish to eat all of the food provided. Un-ionized ammonia levels reached 0.9–1.8 mg/L in the afternoons that probably further reduced fish growth. Fish production at a maximum feeding rate of 112 kg/Ha/day, the highest feeding rate that resulted in an increase in fish production, was 6000 kg/Ha. Cole and Boyd (1986) concluded that, regardless of the amount of aeration applied to fed fish ponds, fish production will ultimately be limited by ammonia accumulation.

At a moderate feeding rate of 53 kg/Ha/day, continuous nighttime aeration (from midnight till dawn) resulted in higher fish production and better food conversion than when the permanently installed aerators were operated on an emergency basis (Lai-fa and Boyd 1988). At stocking and feeding rates more typical of intensive catfish culture, there was no difference between the two methods of aerator deployment (Steeby and Tucker 1988). Thus, intensive catfish production involves prediction of early morning DO levels and operation of dedicated electric paddlewheel aerators as needed. Commercial catfish farmers using intensive aeration report a maximum production of around 6000–8000 kg/Ha in ponds where high rates of water exchange are not possible (Boyd 1998).

Torrens (2005) studied further intensification of catfish culture by using high levels of aeration, continuous water circulation, and water chemistry manipulation to control NO_2 toxicity in small, experimental ponds. Aeration was accomplished with three mechanical paddlewheel aerators and one continuously operated vertical pump circulator in each pond. Each aerator was at a power level of 3.7 kW/Ha for a total power capacity of 11.1 kW/Ha, plus 3.7 kW/Ha for the circulator, for a total power capacity of 14.8 kW/Ha. The paddlewheel aerators were deployed when DO concentrations fell below 5 mg/L in the high oxygen treatment. Nitrite toxicity was controlled by the addition of rock salt to maintain chloride concentrations at 100 mg/L. Partial harvests were conducted twice during the summer.

The total production from the two partial and final complete harvests averaged 13,393 kg/Ha in 2001 and 23,547 kg/Ha in 2002. While these results may not be applicable to commercial-scale ponds, the increase in production possibly due to NO_2 toxicity control should be investigated as a possible means of increased intensity of feed-based static water aquaculture.

Representative static water aquacultures based on artificial diets

Tilapia, principally, *O. niloticus*, and various hybrids, became a global food commodity beginning in the mid-1980s when the industry expanded rapidly with the widespread use of commercial dry diets (Rakocy 2005). Commercialization of tilapia aquaculture would not have been possible without the development of sex-reversal technologies, developed in the 1970s, which allowed for all-male culture. Tilapia can survive DO exposures as low as 0.6 mg/L for short periods, but require concentrations above 2.0 mg/L to avoid significant stress. Chronic exposure below 3.0 mg/L compromises feed utilization, growth, and disease resistance (McGee Undated). Thus, aeration is required for intensive, static water production. Normal yields range from 6000 to 8000 kg/Ha/crop, but yields as high as 10,000 kg/Ha have been reported with water exchange rates as high as 5%–15% of the pond volume daily and aeration at up to 15 kW/Ha (Rakocy 2005). It takes 5 to 8 months to produce a crop of marketable-sized tilapia in tropical climates, so annual production is higher than the per-crop values reported in the United States (Rakocy 2005).

Global tilapia production reached 4.3 million tonnes in 2010 (FAO 2014a,b) (Figure 19.5).

Figure 19.5 Tilapia harvest in China. (Photograph courtesy of Aaron McNevin.)

A recent addition to the global whitefish supply is from the aquaculture of striped catfish, also called Pangasius. Striped catfish, marketed as swai in the West, is a facultative air breather; thus, aeration is not required and production levels of 250,000–300,000 kg/Ha have been reported (Griffiths et al. 2010). This incredible production is accomplished through high rates of water exchange. Nguyan and Oanh (2009) report water exchange rates as high as 40%–60% of pond volume per day. Nearly all striped catfish are currently produced in Vietnam. Ponds are small but deep and situated close to the rivers that supply water necessary for the high flushing rates required for the observed level of production intensity. Striped catfish ponds in Vietnam average 0.32 Ha with a depth of 4.4 m (Nguyan 2013). No information is available on ammonia or nitrite exposures in striped catfish ponds or the tolerance of striped catfish to these metabolites. Apparently the high flushing rates observed for these ponds are sufficient to protect the fish from the toxicity of ammonia and nitrite. Pangasius production in Vietnam was 376,000 tonnes in 2005, but by 2012 had risen to 1.2 million tonnes (FAO 2012a,b). The global Pangasius production was 1.9 million tonnes in 2012. This figure contrasts with 0.14 million tonnes in 2014 for the American channel catfish industry, down from a peak of 0.29 million tonnes in 2004 (USDA 2015).

Most of the global aquaculture production of finfish is carp and 80% of that total is produced in China (Dey et al. 2005). There are more than 20 species of Asian carp, but four species of Chinese carp (silver carp, grass carp, Crucian carp—*Carassius carassius*, and bighead carp), three species of Indian carp (rohu—*Labio rohita*, catla—*Catla catla*, and mrigal—*Cirrhinus cirrhosis*), and common carp are the most important. Of the total world carp production, 23% is silver carp and 22% grass carp (Dey et al. 2005). Asian carp have traditionally been cultured in polyculture systems with different species occupying separate ecological niches. All species have relatively low trophic status so high production is possible with fertilization as the only management input, but recently commercial diets have been used to intensify production (Dey et al. 2005). Asian carp production expanded rapidly following the development of hypophysation in the 1960s, which allowed for availability of large numbers of known-species seed for stocking. Dey et al. (2005) reported that average production levels for carp polyculture are 4474 kg/Ha in China. Global carp production rose from 5.5 million tonnes in 1990 to 16.3 million tonnes in 2001 (Dey et al. 2005), and the latest production reports indicate that global carp production has risen to 23.2 million tonnes (Table 19.1).

Penaeid shrimp, *Peneaus vannamei* and *Peneaus monodon*, dominate global shrimp markets. The most common aquaculture technique involves fertilization and formulated feeds with some aeration and water exchange. This level of intensity results in yields averaging 2000 kg/Ha/crop with two crops per year (Briggs 2006). One method, called the

Table 19.1 Global production of carp in 2012

Species	Production in 2012 (million tonnes)	Reference
Silver carp	4.2	Yang (2005)
Grass carp	5.0	Weimin (2004a)
Common carp	3.8	Peteri (2004)
Bighead carp	2.9	Weimin (2004b)
Crucian carp	2.5	Weimin (2004c)
Rohu	1.6	Jena (2006a)
Catla	2.8	Jena (2006b)
Mrigal	0.4	Ayyappan (2006)
Total	23.2	

bacterial floc system, utilizes low-protein, inexpensive feeds to increase the C:N ratio to 10:1 or greater in order to develop a heterotrophic bacterial system. Shrimp consume the bacterial flocs that develop. These systems require high levels of aeration and produce from 8000 to 50,000 kg/ Ha/crop (Briggs 2006). Global production estimates for *P. vannamei* and *P. monodon* were 3.2 million tonnes (Briggs 2006) and 0.9 million tonnes (Kongkeo 2005), respectively.

Another example of a dramatic increase in production intensity due to the adoption of commercial diets is in the culture of milkfish, *Chanos chanos*. The traditional method uses organic and inorganic fertilizers to grow a complex benthic community as forage for the fish. In the 1980s, commercial feeds were developed and are now used almost exclusively for milkfish culture (Nelson and Marygrace 2007). Global production of milkfish rose from 0.25 million tonnes in 1980 to 0.95 million tonnes in 2012 (Nelson and Marygrace 2007).

SAMPLE PROBLEMS

1. Which pond will have the greatest loss of DO from diffusion, 30°C with 12 mg/L DO at dusk, or 20°C with 14 mg/L at dusk?
2. What will be the overnight (12-hour) decline in DO caused by fish respiration in a 10 Ha pond 1.2 m deep containing 1800 kg/Ha of 300 g catfish? Assume water temperature is 29°C. If 9000 kg of fish are removed and the remaining fish grow so that the pond again contains 1800 kg/Ha, what will the overnight decline in DO caused by fish respiration then be?
3. The pond described in Problem 2 had a Secchi disk visibility of 50 cm in April when the water temperature was 20°C. In July, surface temperatures averaged 29°C and the Secchi disk visibility was reduced to 22 cm. How much did the nighttime decline in DO increase from April to July?

4. Consider a 20-Ha pond containing 2250 kg/Ha of 350 g catfish. The average depth is 1 m, water temperature is 28°C, and the Secchi disk visibility is 20 cm. If the evening DO is 14 mg/L, what will the dawn DO be?

   The next day is cool and cloudy so that the water temperature drops to 24°C and the evening DO is only 8 mg/L. Now what will the morning DO be? Will aeration be necessary?

5. Use the projection method to predict the dawn DO in a pond where the DO at 7:00 p.m. is 11.2 mg/L and the DO at 11:00 p.m. is 8.1 mg/L. Sunrise is at 6:20 a.m.

References

Almendras, J. M. E. 1987. Acute nitrite toxicity and methemoglobinemia in juvenile milkfish *Chanos chanos* Forsskal. *Aquaculture* 61: 33–44.

Andrews, J. W. and Y. Matsuda. 1975. The influence of various culture conditions on the oxygen consumption of channel catfish. *Transactions of the American Fisheries Society* 104: 322–327.

Ayyappan, S. 2006. Cultured Aquatic Species Information Programme, *Cirrhinus mrigala*. FAO Fisheries and Aquaculture Department [online]. Rome, Italy.

Boyd, C. E. 1998. Pond water aeration systems. *Aquacultural Engineering* 18: 9–40.

Boyd, C. E., R. P. Romaire, and E. Johnston. 1979a. Predicting early morning dissolved oxygen concentrations in channel catfish ponds. *Transactions of the American Fisheries Society* 197: 484–492.

Boyd, C. E., J. A. Steeby, and E. W. McCoy. 1979b. Frequency of low dissolved oxygen concentrations in ponds used for commercial culture of channel catfish. *Proceedings of the Annual Conference of the Southeastern Association of Fish and Wildlife Agencies* 33: 591–599.

Briggs, M. 2006. Cultured Aquatic Species Information Programme, *Penaeus vannamei*. FAO Fisheries and Aquaculture Department [online]. Rome, Italy.

Busch, R. L., C. S. Tucker, J. A. Steeby, and J. E. Reames. 1984. An evaluation of three paddlewheel aerators used for emergency aeration of channel catfish ponds. *Aquacultural Engineering* 3: 59–69.

Cole, B. A. and C. E. Boyd. 1986. Feeding rate, water quality and channel catfish production in ponds. *Progressive Fish-Culturist* 48: 25–29.

Dey, M. M., F. J. Paraguas, R. Bhatta, F. Alam, M. Weimin, S. Piumsombun, S. Koeshandrajana, L. T. C. Dung, and N. V. Sang. 2005. Carp production in Asia: Past trends and present status. Pages 6–15 *in* D. J. Penman, M. V. Gupta, and M. M. Dey, editors, *Carp Genetic Resources for Aquaculture in Asia*. WorldFish Center Technical Report 65.

FAO. 2012a. *The State of World Fisheries and Aquaculture 2012*. Food and Agriculture Organization of the United Nations. Rome, Italy.

FAO. 2012b. FAO species fact sheet—*Penaeus monodon* [online].

FAO. 2014a. *The State of World Fisheries and Aquaculture 2014*. Food and Agriculture Organization of the United Nations. Rome, Italy.

FAO. 2014b. Globefish market report—*Tilapia*, December, 2014 (online). Food and Agriculture Organization of the United Nations. Rome, Italy.

Glencross, B., T. T. T. Hien, N. T. Phuong, and T. L. Cam Tu. 2011. A factorial approach to defining the energy and protein requirements of Tra catfish, *Pangasianodon hypothalamus*. *Aquaculture Nutrition* 17: 317–326.

Griffiths, D., P. Van Khanh, and T. Q. Trong. 2010. Cultured aquatic species programme. *Pangasius hypophthalamus*. FAO Fisheries and Aquaculture Department [online]. Rome, Italy.

Jauncey, K. 2000. Nutritional requirement. Pages 327––325 *in* M. C. M. Beveridge and B. J. McAndrew, editors, *Tilapias: Biology and Exploitation*. Kluwer Academic Publishers, London, UK.

Jena, J. K. 2006a. Cultured Aquatic Species Information Programme, *Labeo rohita*. FAO Fisheries and Aquaculture Department [online]. Rome, Italy.

Jena, J. K. 2006b. Cultured Aquatic Species Information Programme, *Catla catla*. FAO Fisheries and Aquaculture Department [online]. Rome, Italy.

Kongkeo, H. 2005. Cultured Aquatic Species Information Programme. *Penaeus monodon*. FAO Fisheries and Aquaculture Department [online]. Rome, Italy.

Lai-fa, Z. and C. E. Boyd. 1988. Nightly aeration to increase the efficiency of channel catfish production. *Progressive Fish-Culturist* 50: 237–242.

Lee, K. J. and S. C. Bai. 1997. Hemoglobin powder as a dietary animal protein source for juvenile Nile tilapia. *Progressive Fish-Culturist* 59: 266–271.

Li, M. H., C. E. Lim, and C. D. Webster. 2006. Feed formulation and manufacture. Pages 517–546 *in* C. E. Lim and C. D. Webster, editors. *Tilapia: Biology, Culture and Nutrition*. Food Products Press, Binghamton, New York.

Lim, C. E. and C. D. Webster. 2006. Nutrient requirements. Pages 469–501 *in* C. E. Lim and C. D. Webster, editors. *Tilapia: Biology, Culture and Nutrition*. Food Products Press, Binghamton, New York.

Losordo, T. M. 1997. Tilapia culture in intensive recirculating systems. Pages 185–211 *in* B. A. Costa-Pierce and J. E. Rakocy, editors, *Tilapia Aquaculture in the Americas*, Vol. 1. The World Aquaculture Society, Baton Rouge, Louisiana.

Lovell, T. 1989. *Nutrition and Feeding of Fish*. Van Nostrand Reinhold, New York, New York.

McGee, M. V. Undated. *Aquaculture of Tilapia and Pangasius; A Comparative Assessment*. Caribe Fisheries, Inc., Lajas, Puerto Rico.

Mezainis, V. E. 1977. Metabolic rates of pond ecosystems under intensive catfish cultivation. MS Thesis, Auburn University, Auburn, Alabama.

Nelson, A. L. and C. Q. Marygrace. 2007. Cultured Aquatic Species Programme, *Chanos chanos*. FAO Fisheries and Aquaculture Department [online]. Rome, Italy.

Nguyan, P. T. and D. T. H. Oanh. 2009. Striped catfish (*Pangasianodon hypophthalamus*) aquaculture in Vietnam: An unprecedented development within a decade. Pages 133–150 *in* S. S. Silva and F. B. Davy, editors, *Success Stories in Asian Aquaculture*. Springer, IDRC, NACA, The Netherlands.

Nguyan, T. P. 2013. On-farm feed management practices for striped catfish (*Pangasianodon hypophthalamus*) Mekong River Delta, Vietnam. Pages 241–267 *in* M. R. Hasan and M. B. New, editors, *On-Farm Feeding and Feed Management in Aquaculture*. FAO Fisheries and Aquaculture Technical Paper No. 583. Rome, Italy.

Perrone, S. J. and T. L. Meade. 1977. Protective effect of chloride on nitrite toxicity to coho salmon, *Oncorhynchus kisutch*. *Journal of the Fisheries Research Board of Canada* 34: 486–492.

Peteri, A. 2004. Cultured Aquatic Species Information Programme, *Cyprinus carpio*. FAO Fisheries and Aquaculture Department [online]. Rome, Italy.

Prather, E. E. and R. T. Lovell. 1971. Effect of vitamin fortification in Auburn No. 2 fish feed. *Proceedings of the Annual Conference of Southeastern Association of Game and Fish Commissioners* 23: 176–178.

Rakocy, J. E. 2005. Cultured Aquatic Species Programme. *Oreochromis niloticus*. FAO Fisheries and Aquaculture Department [online]. Rome, Italy.

Romaire, R. P. and C. E. Boyd. 1978. *Predicting Nighttime Oxygen Depletion in Catfish Ponds*. Bulletin 505, Alabama Agriculture Experiment Station, Auburn University, Auburn, Alabama.

Schroeder, G. L. 1975. Nighttime material balance for oxygen in fish ponds receiving organic wastes. *Bamidgeh* 27: 65–74.

Schwedler, T. E., C. S. Tucker, and M. H. Beleau. 1985. Non-infectious diseases. Pages 497–541 *in* C. S. Tucker, editor, *Channel Catfish Culture*. Elsevier, New York, New York.

Shiau, S. Y., J. L. Chuang, and C. L. Sun. 1987. Inclusion of soybean meal in tilapia (*Oreochromis niloticus* × *O. aureus*) diets at two protein levels. *Aquaculture* 65: 51–261.

Siddiqui, A. Q., M. S. Howlander, and A. A. Adam. 1988. Effects of dietary protein levels on growth, feed conversion and protein utilization in fry and young Nile tilapia, *Oreochromis niloticus*. *Aquaculture* 70: 63–73.

Steeby, J. A. and C. S. Tucker. 1988. *Comparison of Nightly and Emergency Aeration of Channel Catfish Ponds*. Research Report No. 13. Mississippi Agricultural and Forestry Experiment Station, Mississippi State University, Mississippi State, Mississippi.

Tacon, A. G. J., K. Jauncey, A. Falaye, M. Pentah, I. MacGowen, and E. Stafford. 1983. The use of meat and bone meal and hydrolyzed feather meal and soybean meal in practical fry and fingerling diets for *Oreochromis niloticus*. Pages 356–365 *in* J. Fishelton and Z. Yaron, editors, *Proceedings, First International Symposium on Tilapia aquaculture*. Tel Aviv University Press, Israel.

Tomasso, J. R., B. A. Simco, and K. B. Davis. 1979. Chloride inhibition of nitrite induced methemoglobinemia in channel catfish *Ictalurus punctatus*. *Journal of the Fisheries Research Board of Canada* 36: 1141–1144.

Torrens, E. L. 2005. Effect of oxygen management on culture performance of channel catfish in earthen ponds. *North American Journal of Aquaculture* 67: 275–288.

Tucker, C. S., J. A. Hargreaves, and C. E. Boyd. 2008. Better management practices for freshwater pond aquaculture. Pages 151–226 *in* C. S. Tucker and J. A. Hargreaves, editors. *Environmental Best Management Practices for Aquaculture*. Blackwell Publishing, Oxford, UK.

USDA. 1997. United States Department of Agriculture/Animal and Plant Health Inspection Service, Reference to 1996 U.S. Catfish Health and Inspection Practices. Fort Collins, Colorado.

USDA. 2015. *National Agriculture Statistics Service*. Agricultural Statistics Board [online]. Washington, DC.

Wang, R., T. Takeuchi, and T. Watanabe. 1985. Effect of dietary protein levels on growth of *Tilapia nilotica*. *Bulletin of the Japanese Society of Scientific Fisheries* 51: 133–140.

Wedemeyer, G. A. and W. T. Yasutake. 1978. Prevention and treatment of nitrite toxicity in juvenile steelhead trout (*Salmo gairdneri*). *Journal of the Fisheries Research Board of Canada* 35: 822–827.

Weimin, M. 2004a. Cultured Aquatic Species Information Programme, *Cteno-pharyngodon idellus*. FAO Fisheries and Aquaculture Department [online]. Rome, Italy.

Weimin, M. 2004b. Cultured Aquatic Species Information Programme, *Hypoph-thalmichthys nobilis*. FAO Fisheries and Aquaculture Department [online]. Rome, Italy.

Weimin, M. 2004c. Cultured Aquatic Species Information Programme, *Carassius carassius*. FAO Fisheries and Aquaculture Department [online]. Rome, Italy.

Yang, N. 2005. Cultured aquatic species information programme, *Hypophthalmi-chthys molitrix*. FAO Fisheries and Aquaculture Department [online]. Rome, Italy.

Appendix: Solutions to sample problems

Chapter 2

1. 20,000 25 cm rainbow trout weigh 3516 kg. By the end of July the fish will be 26.88 cm in length ($\Delta L = -0.167 + 0.066$ T) and weigh 4367 kg. Anticipated gain is 851 kg so **851 kg** of food is required for July. For August, $\Delta L = 0.691$ mm/day, gain and food requirement are **1091 kg**. For September, $\Delta L = 0.559$ mm/day, gain and food requirement are **1005 kg**.

2. 28.5 cm of growth is required. $\Delta L = 0.625$ mm/day. 285 mm/0.625 mm per day = 456 days + 14 days for swim-up. **Eggs should be delivered on December 15.**

3. 25 cm = 175.8 g
   2 cm = 0.09 g

$$\text{Gain} = 175.7 \,\text{g} \times \frac{1.3\,\text{g food}}{\text{g gain}} \times \frac{\text{tonne}}{10^6} \times \frac{\$430}{\text{tonne}} = \frac{\$0.098}{\text{fish}}$$

4. $\Delta L = \dfrac{19.3\,\text{cm}}{90\,\text{days}} = \dfrac{2.14\,\text{mm}}{\text{day}} \times \dfrac{\text{inch}}{25.4\,\text{mm}} = \dfrac{0.08\,\text{inch}}{\text{day}}$

   $HC = 3 \times C\,\Delta L \times 100 = 3 \times 1.7 \times 0.08 \times 100 = \textbf{40.8}$

5. $K = 0.00444$, $W = 35.5$ g $\times 100,000 = 3552$ kg gain $\times 1.7 = \textbf{6038 kg food}$

6. $HC = 8.55 = 3 \times 1.32 \times \Delta L \times 100$. $\Delta L = 0.0216$ inch/day

Lot #	Wt on Aug 1 (lb)	L on Aug 1 (inch)	L on Aug 15 (inch)	Wt on Aug 15 (lb)
1	0.012	3.11	3.43	0.016
2	0.103	6.35	6.67	0.119
3	1.08	13.88	14.20	1.15

Lot #1: F = 8.55/3.43 = 2.49% BW × 0.016 lb/fish × 22,501 fish = **9 lb food**
Lot #2: F = 8.55/6.67 = **29 lb food**
Lot #3. F = 8.55/14.20 = **355 lb food**

7. ΔL = 0.74 mm/day = 0.029 inch/day
   a. HC = 3 × 1.1 × 0.029 × 100 = **9.55**
       Weight on July 15 = 0.371 lb, Length on July 15 = 11.06 inch
       Length on Aug 1 = 11.5 inch, Weight on Aug 1 = 0.414 lb
       Length on Aug 15 = 11.93 inch, Weight on Aug 15 = 0.462 lb
   b. F = 9.57/11.5 × 215,000 = **741 lbs**
   c. F = 9.57/11.93 = **797 lbs**
8. Weight = 244.6 g, Length = 312 mm, ΔL = 1.48 mm/day
     F = (3 × 1.7 × 1.48)/(312) × 244.6 × 110,000 = 652 kg
9.

$$A: \frac{1.4 \text{ tonne food}}{\text{tonne gain}} \times \frac{\$370}{\text{tonne food}} = \frac{\$518}{\text{tonne gain}}$$

     B: $516/tonne gain

     C: $500/tonne gain

10.
    A: Protein: 380 g/kg × 3.9 C/g = 1482 C
    Fat: 50 g/kg × 8 C/g = 400 C
    CHO: 160 g/kg × 1.6 C/g = 256 C

$$\frac{3850\,C/kg \text{ fish}}{2138\,C/kg \text{ food}} = \frac{1.8\,kg \text{ food}}{kg \text{ gain}} \times \frac{\$0.45}{kg \text{ food}} = \frac{\$0.81}{kg \text{ gain}}$$

   B: 3850/2726 × $0.48 = $0.68/kg gain
   C: 3850/3115 × $0.51 = $0.63/kg gain
   D: 3850/3536 × $0.62 = $0.68/kg gain

Chapter 3

1. $1000\,\text{Ha} \times \dfrac{10,000\,m^2}{Ha} \times \dfrac{1.0\,m}{year} \times \dfrac{year}{31,536,000\,sec} \times 10\% = \dfrac{0.03\,m^3}{sec}$

2. $143,000 \div \left[\dfrac{10,000\,m^2}{Ha} \times 1.0\,cm \times \dfrac{m}{100\,cm} \right] = 8\%$

3. $\left[200\,km^2 \times \dfrac{10^6\,m^2}{km^2} \times \dfrac{0.40\,m}{year} \dfrac{year}{31,536,000\,sec} \right] \div 35\% = \dfrac{7.2\,m^3}{sec}$

4.
$$\frac{0.02\,m^3}{sec} \times \frac{1000\,kg}{m^3} \times 62\,m \times \frac{9.8\,m}{sec^2} \times \frac{1J}{kg(m^2/sec^2)} \times \frac{kW}{1000\,J/sec}$$

$$= \frac{12.15\,kW}{0.8} \times \frac{8760\,hour}{year} \times \frac{\$0.15}{kWh} = \frac{\$19,960}{year}$$

5. $100\,kW \times \dfrac{1000\,J/sec}{kW} \times \dfrac{kg(m^2/sec^2)}{J} \times \dfrac{9.8\,m}{sec^2} \times \dfrac{0.075\,m^2}{sec}$

$$\times \frac{m^3}{1000\,kg} = 73.5\,m$$

The drawdown is $73.5 - 35 = \textbf{38.5 m}$

6. $Q = VA$, $A = Q/V = 0.35/0.5 = 0.7\,m^2$

The area of a trapezoid is $A = hb + h(c - b/2)$ where h, b, and c are lengths of the height, base, and crown, respectively. For a trapezoid with a 2:1 side slope, $c - b = 4\,h$. If we let $b = 2\,h$, the cross-sectional surface area to flow is $h(2\,h) + h(2\,h) = 07$ and $h = 0.0418$. **The dimensions of the channel are 0.418:0.836:2.51 m (h:b:c).**

7. Conveyance Method:

$$K = \left[6.304 \left(\frac{2\log D}{e} + 1.14 \right) D^{2.5} \right]$$

Try 1-inch pipe:
$L = 197 + 2(1/12 \times 15) + (1/12 \times 20) + (1/12 \times 17) = 202.6$

$$K = \left[6.304 \left(2\log \frac{1}{10^{-6}} + 1.14 \right) \left(\frac{1}{12} \right)^{2.5} \right]$$

$$Q = 0.166 \sqrt{\frac{1.64}{202.6}} = 0.015\,cfs = 25\,L/min$$

Try 2-inch pipe: $K = 0.911$, $Q = 139\,L/min$
Try 2.5-inch pipe: $K = 1.64$, $Q = 251\,L/min$

Select the 2.5-inch pipe.
Manning Equation:

$$Q = \frac{0.3116}{0.018} D^{2.67} S^{0.5}$$

$$\frac{0.0033\,m^2}{sec} = \frac{0.3116}{0.018} D^{2.67} \left(\frac{0.5}{60} \right)^{0.5}$$

$D = 0.1\,m = 10\,cm.$

8. $V = \dfrac{1}{0.015}(0.15^{2/3})\left(\dfrac{1}{1000}\right)^{1/3} = \dfrac{1.9\,\text{m}}{\text{sec}}$

$Q = 0.18 \times 1.9 = \dfrac{0.3385\,\text{m}^3}{\text{sec}} = \dfrac{20{,}314\,\text{L}}{\text{sec}}$

9. $N_s = \dfrac{1800\sqrt{0.0033}}{70^{0.75}} = \textbf{4.29 (metric)}$

$N_s = \dfrac{1800\sqrt{52}}{230^{0.75}} = \textbf{220 (English)}$

$\dfrac{0.0033\,\text{m}^3}{\text{sec}} \times \dfrac{1000\,\text{kg}}{\text{m}^3} \times 70\,\text{m} \times \dfrac{9.8\,\text{m}}{\text{sec}^2} \times \dfrac{\text{J}}{\text{kg}\,\text{m}^2/\text{sec}^2} \times \dfrac{\text{kW}}{1000\,\text{J}/\text{sec}} = \textbf{2.26 kW}$

10. $Q = 1.37\,(0.2)^{2.5} = \textbf{0.025 m}^3\textbf{/sec}$

Chapter 4

1. $2400\,\text{L/min} = 2.4\,\text{m}^3/\text{min} \times 15\,\text{min} = 36\,\text{m}^3 = \text{Volume}$
   $30x \times 3x \times x = 36$, $x = \textbf{0.74 m} = \text{Depth}$, $30x = \textbf{22 m} = \text{Length}$, $3x = \textbf{2.2 m} = \text{Width}$
2. $Q = V\,A$

$$V = \dfrac{0.04\,\text{m}^3/\text{sec}}{1.63\,\text{m}^2} = 0.024\,\text{m/sec} = 2.46\,\text{cm/sec}$$

$$A = \dfrac{0.04}{0.033} = 1.21\,\text{m}^2$$

If Depth = 0.74 m, Width = 1.64 m, and Length = 29.7 m.

$A = 1.21\,\text{m}^2 = 3x$ where $x = \text{Depth}$. Depth = 0.64 m, Width = 1.91 m, and Length = 19.1 m. **Volume is now 23.3 m³ and R = 6.17.**

3. $\dfrac{15.77\,\text{m}^3}{6\,\text{min}} \times \dfrac{1000\,\text{L}}{\text{m}^3} = \dfrac{2628\,\text{L}}{\text{min}}$

$$V = \dfrac{0.044}{2.16} = \dfrac{0.020\,\text{m}}{\text{sec}}$$

$$Q = 0.033 \times 2.16 = \dfrac{0.071\,\text{m}^3}{\text{sec}}$$

Increase flow to 4260 L/min or reduce depth to 0.55 m.

4. $163\,L \times \dfrac{\min}{45\,L} = \textbf{3.6 min}$

$A = \pi r^2 = 0.246\ m^2$

$$V = \dfrac{0.0075}{0.246} = \textbf{0.003 m/sec}$$

5. $V = (\pi)(0.5^2)(3) = \dfrac{2.36\,m^3}{12\,\min} \times \dfrac{1000\,L}{m^3} = \textbf{197 L/min}$

6. $\dfrac{30\,L}{sec} \times \dfrac{60\,sec}{\min} \times \dfrac{m^3}{1000\,L} \times \dfrac{30\,\min}{exchange} = 54\,m^3 = \text{Volume}$

Set Depth at **1.0 m**, $54 = \pi r^2 \times 1$, $r = 4.15$ m, Diameter = **8.3 m**

7. Circumference = 26 m
   Depth = 1 m + 0.5 m freeboard, so walls are $26 \times 1.5 \times 0.2 = 7.8\ m^3$. Floor is $54\ m^2 \times 0.2$ m thick $= 10.8\ m^3$. Total volume of concrete required is $18.6\ m^3 \times \$200 = \textbf{\$3720.}$

8. Raceway with 5 exchanges per hour has a volume of

$$\dfrac{30\,L}{sec} \times \dfrac{60\,sec}{\min} \times \dfrac{m^3}{1000\,L} \dfrac{12\,\min}{exchange} = 21.6\,m^3$$

If dimensions are 30:3:1, Depth = 0.62 m, Width = 1.86 m, Length = 18.6 m. Leave 0.5 m freeboard so walls are $18.6 \times 1.12 \times 0.2 = 4.1$ 7 $m^3 \times 2 = 8.33\ m^3$. Floor is $18.6 \times 1.86 \times 0.2 = 6.92\ m^3$. Ends are open so total volume of concrete required is $15.25\ m^3 \times \$200 = \textbf{\$3050.}$ Note: Side by side raceways share a common center wall so the material savings for raceway construction are greater than are shown here.

9. Volume of circular pond is 54 m^3 and volume of raceway is 21.6 m^3 so fish densities in the circular pond are **39 kg/m³ and 97 kg/m³** in the raceway.

10. 40 minutes = **1.5 exchanges per hour.**

Chapter 5

1. $0.8705 \times \dfrac{44,000\,mg}{mole} \times \dfrac{mole}{24\,L} \times 0.000407 = \textbf{0.65 mg/L}$

2.

  a. $Ce = 14.161 - 0.3943(20) + 0.0077147(20^2) - 0.0000646(20^3)$

$$\times \frac{760}{760 + (1500/32.8)}$$

  $= 8.34\,mg/L$

  b. **7.14 mg/L**

  c. **10.55 mg/L**

  d. **8.84 mg/L**

3.

  a. $Ce = 10.49\ mg/L$, $PO_2a = 152.92$, $DO = 5.0$

$$\frac{5.0}{10.49} = \frac{PO_2w}{159.92} = 72.89\,mm\,Hg$$

  % saturation $= 5/10.49 = $ **47.7%**

  b. $\dfrac{12.0}{7.13} = \dfrac{PO_2w}{154.1} = $ **259.36 mm Hg, 168.3% saturation**

  c. $\dfrac{11.47 \times 2.3}{10.14} = \dfrac{PO_2w}{140.73} = $ **366.14 mm Hg, 260.2% saturation**

  d. $\dfrac{10.96}{8.55} = \dfrac{PO_2w}{153.95} = $ **197.34 mm Hg, 128.2% saturation**

4. $0.5 = \dfrac{1}{1 + E/7600}$ $E = $ **7600 m above sea level**

  (24,928 ft above sea level)

5. $7.75 - 17[0.0841 - 0.00256(28) + 0.0000374(28^2)] = $ **7.05 mg/L**

6. $DO = 9.76 - 1.89 = 7.87\ mg/L$

$$\frac{7.87\,mg}{L} \times \frac{1000\,L}{m^3} \times \frac{10^9\,m^3}{km^3}\,\frac{kg}{10^6\,kg} = 7{,}870{,}000\,kg$$

7. $8.11/0.20946 = $ **38.7 mg/L**

8. $P = 745.05\ mm\ Hg$, $PO_2a = 156.06\ mm\ Hg$, Pw at 10% RH $= 0.92$ mm Hg and at 90% RH $= 8.28$ mm Hg.

$$PO_2w = \frac{5.0}{10.25} \times (712.88 - 0.92) \times 0.20946 = 72.74\,mm\,Hg\ \text{at 10% RH}$$

$$PO_2w = \frac{5.0}{10.25} \times (712.88 - 8.28) \times 0.20946 = 71.99\,mm\,Hg\ \text{at 90% RH}$$

Difference is 0.75 mm Hg

9.

  a. $\dfrac{0.20946\,(726.54 - 4.6)}{0.20946\,(726.54 - 0)} = $ **0.6%**

b. $\dfrac{(5/10.22) \times 0.20946\,(713.09 - 31.8)}{\dfrac{5}{10.22} \times 0.20946\,(713.09 - 0)} = \dfrac{69.82}{73.07} = \mathbf{4.4\%}$

c. $\dfrac{760 - 31.8\,(0.20946)}{760 - 0\,(0.20946)} = \mathbf{4.2\%}$

d. $\dfrac{67.83}{71.85} = \mathbf{5.6\%}$

10. $\dfrac{3.5}{7.75} \times 159.2 = 71.9\,\text{mm Hg} = \dfrac{DO}{11.47} \times 159.2 = \mathbf{5.18\,mg/L}$

Chapter 6

1.

a. $O_2 = (7.2 \times 10^{-7})(46.4^{3.2})(0.225^{-0.194}) = 0.207\ \text{mg/100 mg day}$

$$\frac{0.207\,\text{mg}}{100\,\text{mg day}} \times \frac{10^4\,(100\,\text{mg})}{\text{kg}} \times \frac{\text{day}}{24\,\text{hr}} = \mathbf{86.2\,mg/kg\ hour}$$

b. $O_2 = (1.9 \times 10^{-6})(46.4^{3.13})(0.26^{-0.138}) = 0.376$ mg/100 mg day = **156.7 mg/kg hour**

c. $O_2 = (4.9 \times 10^{-5})(51.8^{2.12})(0.093^{-0.194}) = 0.334$ mg/100 mg day = **139.2 mg/kg hour**

d. $O_2 = (3.05 \times 10^{-4})(59^{1.855})(0.381^{-0.138}) = 0.672$ mg/100 mg day = **280 mg/kg hour**

2. $OD = (75)(117.9^{-0.196})(10^{0.055 \times 8}) = \mathbf{81.1\ mg/kg\ hour}$

$OD = (249)(181.4^{-0.142})(10^{0.024 \times 15}) = \mathbf{272.6\ mg/kg\ hour}$

3. DO available = 7.75 − 2.43 = 5.32 mg/L.

$\log\ O_2 = -0.999 - 0.000957(50) + 0.0000006(50^2) + 0.0327(28) - 0.0000087(28^2) + 0.0000003(50 \times 28) = -0136.$ $O_2 = 0.731$ mg/g hour.

$$\frac{\text{g hour}}{0.731\,\text{mg}} \times 5000\,\text{g} \times \frac{5.32\,\text{mg}}{\text{L}} \times \frac{1000\,\text{L}}{\text{m}^3} \times 0.234\,\text{m}^3 = 0.34\ \text{hour} = \mathbf{20\,min}$$

4. $Oc = \dfrac{3 \times 1.5 \times 1.49}{184.5} \times 9155.23 = 332\,\text{mg/(kg hour)} = \mathbf{43\,min}$

5. 6.5: $Oc = \dfrac{3 \times 1.5 \times 1.45}{334.4} \times 9155.23 = 178.6\,\dfrac{\text{mg}}{\text{kg hour}}$

$$\frac{178.6\,\text{mg}}{\text{kg hour}} \times \frac{\text{kg}}{10^6\,\text{mg}} \times 1350\,\text{kg} \times \frac{\text{kW h}}{0.6\,\text{kg}} = \mathbf{0.4\,kW}$$

6.6:   log $O_2 = -0.353$, $O_2 = 0.443$ mg/g   hour = 443 mg/kg   hour. Required power is **1.0 kW**.

6. 6.5 : $Oc = \dfrac{3 \times 1.5 \times 1.48}{396.4} \times 9155.23 = \mathbf{153.8\,mg/kg\,hour}$

6.6: log $O_2 = -0.382$, $O_2 = 0.418$ mg/g hour = **418 mg/kg hour**

7. $\dfrac{5.0}{10.51} = \dfrac{PO_2 w}{153.17} = 72.86$ mm Hg

$$\dfrac{DO}{7.81} = \dfrac{72.86}{158.16} = \mathbf{3.6\,mg/L}$$

8. Location 1: Ce = 10.40, 75 mm Hg = 5.03 mg/L. Available DO = **5.37 mg/L**

   Location 2: Ce = 7.34, 75 mm Hg = 3.55 mg/L. Available DO = **3.79 mg/L**

9. $Oc = \dfrac{3 \times 1.5 \times 0.625}{80} \times 9155.23 = 321.9$ mg/kg hour

$$\dfrac{321\,mg}{kg\,hour} \times 2840.4\,kg \times \dfrac{min}{2300\,L} \times \dfrac{hour}{60\,min} = 6.63\,mg/L$$

DO = 9.76 − 6.63 = 3.13 mg/L

$$\dfrac{3.13}{9.76} \times 149.33 = \mathbf{47.9\,mm\,Hg}$$

10. 75 mm Hg appears to be a reasonable compromise between fish health and fish production when liquid oxygen is not available to maintain DO levels at 100 mm Hg.

Chapter 7

1. Use 10°C, 254 mm trout and P = 760 mm Hg

Equation 7.1: $CC = \dfrac{(8.74 - 5.14)(0.0545)}{0.00815} = 38.65\,lb/gpm$

$= \mathbf{4.64\,kg/L/min}$ **First pond**

$CC = \dfrac{(8.74 - 5.14)(0.0545)}{0.00815} = 24.07\,lb/gpm$

$= \mathbf{2.89\,kg/L/min}$ **Second and third ponds**

Equation 7.2: CC = 1.8 × 10 = 18 lb/gpm = **2.16 kg/L/min First pond**
CC = 1.44 × 10 = 14.4 lb/gpm = **1.73 kg/L/min.**

2. $CC = \dfrac{(7.75 - 3.60)(0.0545)}{(3 \times 1.5 \times 1.49)/368} = 12.4\,\text{lb/gpm} = 1.49\,\text{kg/L/min}$

$$\dfrac{1.49\,\text{kg}}{\text{L/min}} \times \dfrac{\text{fish}}{0.4\,\text{kg}} \times \dfrac{1000\,\text{L}}{\text{m}^3} \times \dfrac{30\,\text{m}^3}{\text{sec}} \times \dfrac{60\,\text{sec}}{\text{min}} = 6,705,000\,\text{fish}$$

$$2,682,000\,\text{kg} \times \dfrac{\text{m}^3}{160\,\text{kg}} = \mathbf{16,763\,m^3 \ of \ rearing \ volume \ required}$$

3. Elevation is 300 m above sea level

$$CC = \dfrac{(10.76 - 5.27)(0.0545)}{0.0096} = 31.2\,\text{lb/gpm} = 3.74\,\text{kg/L/min}$$

$$\dfrac{3.74\,\text{kg}}{\text{L/min}} \times 10,000\,\text{L/min} = 37,400\,\text{kg} \times \dfrac{\text{fish}}{0.037\,\text{kg}} = 998,760\,\text{fish}$$

$$37,400\,\text{kg} \times \dfrac{\text{m}^3}{160\,\text{kg}} \times \dfrac{1.2\,\text{m of length}}{\text{m}^3} = \mathbf{194.8\,m \ of \ length \ required}$$

4. $CC = \dfrac{(8.63 - 4.60)(0.0545)}{0.0148} = \dfrac{14.84\,\text{lb}}{\text{gpm}} = \dfrac{1.78\,\text{kg}}{\text{L/min}} = 1200\,\text{L/min}$

5. Ce = 7.75–0.42 = 7.33 mg/L,   75 mm Hg = 3.45 mg/L, 3.88 mg/L available.
   Use Equation 6.7[30]: log OC = 3.34 − 0.586 (log wt)
   log OC = 1.785
   OC = 60.95 mg/kg hour

$$100,000\,\text{kg} = \dfrac{60.95\,\text{mg}}{\text{kg hour}} \times \dfrac{\text{L}}{3.88\,\text{mg}} \times \dfrac{\text{hour}}{60\,\text{min}} = \mathbf{26,182\,L/min}$$

6. Set T = 10°C, P = 1.0 atm, fish sizes 100, 200, and 400 g

Fish size	CC in kg/L/min			
	7.1	7.2	7.4	7.5
100 g	3.54	1.77	1.71	3.20
200 g	4.47	2.23	1.88	6.49
400 g	6.82	2.81	2.07	7.45

7. $CC = \dfrac{(10.67 - 5.03)(0.0545)}{(3 \times 1.5 \times 0.365)/200} = 37.4\,\text{lb/gpm} = 4.49\,\text{kg/L/min}$

   $\times\ 1700\,\text{L/min} = \mathbf{7633\,kg}$

8. $\dfrac{3850\,\text{C}}{\text{kg fish}} \times \dfrac{\text{kg food}}{2600\,\text{C}} = \dfrac{1.48\,\text{kg food}}{\text{kg gain}}$

$$\dfrac{3850}{3300} = \dfrac{1.17\,\text{kg food}}{\text{kg gain}}$$

For 400 g trout at 10°C, low Calorie diet,

$$CC = \dfrac{(10.92 - 5.14)(0.0545)}{(3 \times 1.48 \times 0.493)/330} = 47.49\,\text{lb/gpm} = \mathbf{5.7\,kg/L/min}$$

And for the high Calorie diet,

$$CC = \dfrac{(10.92 - 5.14)(0.0545)}{(3 \times 1.17 \times 0.493)/330} = 60.07\,\text{lb/gpm} = \mathbf{7.2\,kg/L/min}$$

9. $0.942\,\text{m}^3 \times \dfrac{200\,\text{kg}}{\text{m}^3} \times \dfrac{\text{fish}}{0.342\,\text{kg}} = \mathbf{551\ fish}$

Chapter 8

1. Ce = 9.67 mg/L, 75 mm Hg = 4.92 mg/L
   DO added by simple weirs = 0.093 (9.67 − 4.92) = 0.44 mg/L × 2 = 0.88 mg/L
   Total DO available = (9.67 − 4.92) + (0.88) = 5.63 mg/L
   DO added by lattices = 0.34 (9.67 − 4.92) = 1.62 mg/L × 2 = 3.23 mg/L
   Total DO available = 9.67 − 4.92) + (3.23) = 7.98 mg/L

With weirs, $CC = \dfrac{(5.63)(0.0545)}{(3 \times 1.5 \times 0.625)/250} = 27.28\,\text{lb/gpm} = 3.28\,\text{kg/L/min}$

$\times\ 1800\,\text{L/min}$

$= 5904\,\text{kg} \times \dfrac{\text{fish}}{0.173\,\text{kg}} = 34{,}127\ \text{fish}$

With lattices, $CC = \dfrac{(7.98)(0.0545)}{0.01125} = 38.66\,\text{lb/gpm} = 4.64\,\text{kg/L/min}$
$\times 1800\,\text{L/min}$

$$= 8352\,\text{kg} \times \dfrac{\text{fish}}{0.173\,\text{kg}} = \textbf{48,277 fish}$$

Weight difference $= 8352 - 5904 = 2448\text{ kg} \times \$5.00/\text{kg} = \textbf{\$12,240}$

2. $RT = \dfrac{1.5\,(10.71 - 5.0)(1.025^{10-20})(0.85)}{8.84} = 0.64\,\text{kg/kWh}$

$$\dfrac{0.64\,\text{kg}}{\text{kWh}} \times \dfrac{10^6\,\text{mg}}{\text{kg}} \times 1.5\,\text{kW} \times \dfrac{\text{min}}{12,000\,\text{L}} \times \dfrac{\text{hour}}{60\,\text{min}} = 1.33\,\text{mg/L}$$

$$\text{Second unit}: RT = \dfrac{1.5\,(10.71 - 6.33)(1.025^{-10})(0.85)}{8.84}$$
$$= 0.49\,\text{kg/kWh} = 1.02\,\text{mg/L}$$

DO below first aerator = 6.33 mg/L, below second unit = 7.35 mg/L.

3. Correct size to restore DO to 90% saturation (9.64 mg/L),

$$\dfrac{\text{kWh}}{0.64\,\text{kg}} \times \dfrac{\text{kg}}{10^6\,\text{mg}} \times \dfrac{4.64\,\text{mg}}{\text{L}} \times \dfrac{12,000\,\text{L}}{\text{min}} \times \dfrac{60\,\text{min}}{\text{hour}} = \textbf{5.22 kW}$$

4. $\dfrac{4.64\,\text{mg}}{\text{L}} \times \dfrac{12,000\,\text{L}}{\text{min}} \times \dfrac{\text{kg}}{10^6\,\text{mg}} \times \dfrac{525,600\,\text{min}}{\text{year}} = \textbf{29,265 kg}$

5. Electric : $5.22\,\text{kW} \times \dfrac{8760\,\text{hour}}{\text{year}} \times \dfrac{\$0.15}{\text{kWh}} = \textbf{\$6859/year}$

LOX: 29,265 kg/year at 90% absorption = 32,517 kg/year $\times$ \$0.109/ kg = \$3544 + \$500/month $\times$ 12 months = \$6000. Total LOX cost is **\$9544/year.**

6. $Ce_{DO} = 10.77$ mg/L, 25% saturation = 2.69 mg/L, target DO = 9.69 mg/L, 7.0 mg/L required. If RS = 1.5 kg/kWh, RT = 0.89 kg/kWh

$$\dfrac{\text{kWh}}{0.89\,\text{kg}} \times \dfrac{\text{kg}}{10^6\,\text{mg}} \times \dfrac{7.0\,\text{mg}}{\text{L}} \times \dfrac{2000\,\text{L}}{\text{min}} \times \dfrac{60\,\text{min}}{\text{hour}} = 0.94\,\text{kW required for DO}$$

$Ce_N = 17.83$ mg/L, 125% saturation = 22.29 mg/L, target DN = 110% saturation = 19.61 mg/L, required removal is 22.29 − 19.61 = 2.67 mg/L.

$$RT_{DN} = 1.51 \times 1.5 \times \frac{(22.29 - 17.83)(1.025^{-10})(0.85)}{14.88} = 0.46 \, \text{kg/kWh}$$

$$\frac{kWh}{0.46 \, \text{kg}} \times \frac{kg}{10^6 \, \text{mg}} \times \frac{2.67 \, \text{mg}}{L} \times \frac{2000 \, \text{L}}{\min} \times \frac{60 \, \text{min}}{\text{hour}} = 0.32 \, \text{kW required for DN}$$

Need the 0.94 kW unit to accomplish both goals.

7. Oxygenation will force out excess DN.
   10.77 − 2.69 = 8.08 mg/L DO required.

$$\frac{8.08 \, \text{mg}}{L} \times \frac{2000 \, \text{L}}{\min} \times \frac{525,600 \, \text{min}}{\text{year}} \times \frac{kg}{10^6 \, \text{mg}} \times \frac{\$0.109}{\text{kg}} \div 0.9 \text{ absorption}$$

= $1029 for LOX

$$\text{Electric cost}: 0.94 \, \text{kW} \times \frac{8760 \, \text{hour}}{\text{year}} \times \frac{\$0.15}{\text{kWh}} = \textbf{\$1,235/year for electric}$$

Chapter 9

1. A = 56 P = 56 (.25) = $\dfrac{14.0 \, \text{g TAN}}{\text{kg food}} \times \dfrac{120 \, \text{kg food}}{\text{day}}$

   $= \dfrac{1680 \, \text{g TAN}}{10,000 \, \text{m}^3} \times \dfrac{10^3 \, \text{mg}}{g} \times \dfrac{\text{m}^3}{1000 \, \text{L}}$

   = 0.17 mg/L + 1.0 mg/L = **1.17 mg / L**

2. $f_{6.5} = \dfrac{1}{10^{9.155-6.5} + 1} = 0.0022$

   TAN = NH_3/f = 0.02/0.0022 = 0.91 mg/L
   $f_{9.5}$ = 0.689 × 0.91 = **0.63 mg/L NH_3**

3. $f = \dfrac{1}{10^{9.340-7.2} + 1} = 0.0079$

   0.016/0.0079 = 2.03 mg/L = maximum allowable TAN
   A = 56 P = 25.2 g TAN/kg food

   $$\frac{\text{kg fish}}{0.00839 \, \text{kg food}} \times \frac{\text{kg food}}{25.2 \, \text{g TAN}} \times \frac{g}{1000 \, \text{mg}} \times \frac{2.03 \, \text{mg}}{L} \times \frac{500 \, \text{L}}{\min}$$

   $$\times \frac{1440 \, \text{min}}{\text{day}} = \textbf{6913 \, kg fish}$$

4. $CC = \dfrac{(9.83 - 4.74)(0.0545)}{0.015} = 18.49 \, lb/gpm = 2.22 \, kg/L/min \times 3$

$= 6.66 \, kg/L/min$

$A = \dfrac{28.32 \, g \, TAN}{kg \, food} \times \dfrac{0.015 \, kg \, food}{kg \, fish} \times \dfrac{6.66 \, kg \, fish}{L/min} \times \dfrac{day}{1440 \, min} \times \dfrac{1000 \, mg}{g}$

$= 1.96 \, mg/L$

$=$ maximum allowable TAN

$f = 0.0064$ when $NH_3 = 0.0125$

$$0.0064 = \dfrac{1}{10^{9.766 - pH} + 1}$$

Solve for pH. **pH = 7.57**

5. $f = \dfrac{1}{10^{9.095 - 9.5} + 1} = 0.717$

$CC = \dfrac{(6.29 - 3.55)(0.0545)}{0.03} = 4.98 \, lb/gpm = 0.60 \, kg/L/min$

$\dfrac{17.92 \, g \, TAN}{kg \, food} \times \dfrac{0.03 \, kg \, food}{kg \, fish} \times \dfrac{0.60 \, kg \, fish}{L/min} \times \dfrac{day}{1440 \, min} \times \dfrac{1000 \, mg}{g}$

$= 0.224 \, mg \, TAN/L$

$f = 0.717$, so $NH_3 = 0.161$ mg/L. If this is lost to the atmosphere, 0.063 mg/L will remain and the influent concentration of NH_3 to the second set of ponds will be **0.048 mg/L**.

6. $CC = \dfrac{(9.25 - 4.36)(0.0545)}{0.015} = 17.77 \, lb/gpm = 2.13 \, kg/L/min \times 10$

$= 21.34 \, kg/L/min$

$\dfrac{26.88 \, g \, TAN}{kg \, food} \times \dfrac{0.015 \, kg \, food}{kg \, fish} \times \dfrac{21.34 \, kg \, fish}{L/min} \times \dfrac{1000 \, mg}{g} \times \dfrac{day}{1440 \, min}$

$= 5.98 \, mg/L \, TAN$

$f = 0.00196 \times 5.98 = 0.0117$ mg/L NH_3. This is less than the chronic toxicity level of 0.0125 mg/L so **biofilters are not required.**

7. $67.5\,\text{m}^3 \times \dfrac{176\,\text{kg}}{\text{m}^3} \times \dfrac{0.01\,\text{kg food}}{\text{kg fish}} \times \dfrac{23.52\,\text{g TAN}}{\text{kg food}} \times \dfrac{0.0058\,\text{g NH}_3}{\text{g TAN}}$

$\times \dfrac{1000\,\text{mg}}{\text{g}} \times \dfrac{\text{sec}}{113\,\text{L}} \times \dfrac{\text{day}}{86,400\,\text{sec}}$

$= 0.0017\,\text{mg/L NH}_3$

8. Compile the following table tabulating cumulative oxygen consumption (COC) and specific growth rate (SGR).

Serial unit	Cumulative oxygen consumption	Specific growth rate
1	2.5	0.789
2	5.1	0.819
3	7.5	0.618
4	9.9	0.398
5	12.4	0.204

Regress cumulative oxygen consumption on specific growth rate.

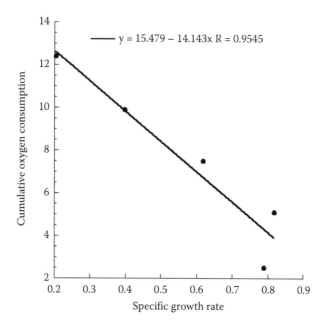

The coefficient of correlation is 0.95, the slope is −14.14 and the intercept is 15.48. The equation for the line is COC = 15.48 − 14.14 (SGR).

The SGR in the first container is 0.789 and half of this is 0.395. Thus the $ECOC_{50} = 15.48 - 14.15 (0.395) = 9.89$ mg/L.

Serial unit	Cumulative fish load	Cumulative oxygen consumption
1	8.5	2.5
2	17.2	5.1
3	24.6	7.5
4	30.8	9.9
5	36.1	12.4

To estimate carrying capacity, regress COC against cumulative fish load using the following data.

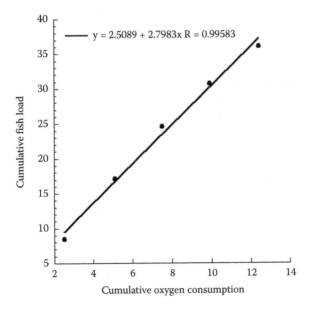

$y = 2.5089 + 2.7983x \quad R = 0.99583$

Cumulative fish load

Cumulative oxygen consumption

The resulting equation is Fish Load = 2.52 + 2.80 (COC) and carrying capacity is the fish load resulting from a COC equal to the $ECOC_{50}$. CC = 2.51 + 2.80 (9.89) = 30.2 kg/8 L/min = 3.78 kg/L/min. Carrying capacity with respect to oxygen is

$$CC = \frac{(9.83 - 5.23)(0.0545)}{0.021} = 9.94 \, lb/gpm = 1.19 \, kg/L/min$$

Metabolite considerations allow for 3.78/1.19 = 3.18 water uses so the hatchery for which the PCA was conducted should be designed with rearing units placed three in a series.

9. $\dfrac{\text{kg fish}}{0.021\,\text{kg food}} \times \dfrac{\text{kg food}}{23.52\,\text{g TAN}} \times \dfrac{\text{g TAN}}{0.0063\,\text{g NH}_3} \times \dfrac{\text{g}}{1000\,\text{mg}}$

$\times \dfrac{0.016\,\text{mg}}{\text{L}} \times \dfrac{1440\,\text{min}}{\text{day}} = \textbf{7.4\,kg/L/min}$

10. $\dfrac{\text{kg fish}}{0.0113\,\text{kg food}} \times \dfrac{\text{kg food}}{23.52\,\text{g TAN}} \times \dfrac{\text{g}}{1000\,\text{mg}} \times \dfrac{2\,\text{mg}}{\text{L}} = \textbf{10.84\,kg/L/min}$

Chapter 10

1. The commonly cited value of 90% solids removal in the QZ was based on a misinterpretation of a study conducted in Lamar, Pennsylvania (Mudrak 1981) and the report from an environmental consulting firm studying hatchery discharges in Idaho (JRB Associates 1984). At one hatchery in Idaho, effluent TSS concentrations were measured in the year prior to installation of QZs and these values were compared to levels measured in the following year after installation of the QZs. At two other hatcheries, effluent concentrations between screened and unscreened raceways were compared during the same year. Fish loads and resulting feeding rates were not reported for these comparisons and the actual weights of solids collected in the QZs were not compared to actual amounts lost over the dams. Sampling procedures and frequencies for TSS concentrations in the QZ-treated effluents were not reported.

2. $\dfrac{1000\,\text{L}}{\text{min}} \times \dfrac{\text{min}}{60\,\text{sec}} \times \dfrac{1000\,\text{cm}^3}{\text{L}} = \dfrac{16{,}667\,\text{cm}^3}{\text{sec}}$

$$\text{OFR (cm/sec)} = \dfrac{Q\,(\text{cm}^3/\text{sec})}{A\,(\text{cm}^2)}$$

$$\text{OFR of }0.5\,\dfrac{\text{cm}}{\text{sec}} = \dfrac{16{,}667\,\text{cm}^3/\text{sec}}{A}\,,\ A = 33{,}334\,\text{cm}^2 \times \dfrac{\text{m}^2}{10{,}000\,\text{cm}^2} = 3.3\,\text{m}^2.$$

$$W = 2.5\,\text{m so } \mathbf{L = 1.3\,m}$$

If OFR = 1.0, A = 1.7 m² and **L = 0.68 m**

3. $A = 2.5 \text{ m} \times 1.2 \text{ m} = 3 \text{ m}^2 = 30,000 \text{ cm}^2$
   $0.5 = Q/30,000, Q = 15,000 \text{ cm}^3/\text{sec} = \textbf{900 L/min}$
   $1.0 = Q/30,000, Q = 30,000 \text{ cm}^3/\text{sec} = \textbf{1800 L/min}$

4. Diet contains 1% P, 69% of which is contained in the discharge. An efficient QZ removes 12% of the solids and the solids contain 57% of total P. Thus,

$$100,000 \text{ kg fish} \times \frac{0.01 \text{ kg feed}}{\text{kg fish}} \times \frac{0.01 \text{ kg P}}{\text{kg feed}} \times 69\% = \frac{6.9 \text{ kg P}}{\text{day}} \text{ in QZ influent}$$

$$\frac{6.9 \text{ kg P}}{\text{day}} - (0.12 \times 0.57) = \frac{6.83 \text{ kg P}}{\text{day}} \text{ in QZ effluent}$$

$$\frac{6.83 \text{ kg P}}{\text{day}} \times \frac{10^6 \text{ mg}}{\text{kg}} \times \frac{\text{day}}{1440 \text{ min}} \times \frac{\text{min}}{30,000 \text{ L}} = \textbf{0.16 mg/L}$$

5. If the microscreens remove 80% of TSS and 57% of TP is bound to the solids, $0.8 \times 0.57 = 45.6\%$ of the TP in the QZ discharge will be captured by the microscreen and the TP concentration in the discharge will be reduced to **0.087 mg/L.**

6. 85% of TN in the hatchery discharge is ammonia. If the diet contains 42% protein, $A = 56P = 23.5$ g TAN/kg food $\times$ 1000 kg food/day 30,000 L/min $= 0.54$ mg/L TAN $= 0.63$ mg/L TN/L. If 19% of TN is in the solids fraction and the QZ removes 12% of the solids, 0.63 mg/L $- (0.19 \times 0.12) = \textbf{0.61 mg TN/L}$ in QZ discharge.

7. $0.8 \times 0.19 = 15.2\%$ of TN in discharge captured by microscreen and TN in discharge is reduced to **0.52 mg TN/L.**

8. QZ removes 6.8% of TP and fish retain 31% so 1200 kg P/year in discharge $= 3175$ kg P/year produced by the fish, or 8.7 kg TP/day. If the diet contains 1% P and the feeding rate is 1% BW per day, the average maximum fish biomass is

$$\frac{8.7 \text{ kg P}}{\text{day}} \times \frac{\text{kg food}}{0.01 \text{ kg P}} \times \frac{\text{kg fish}}{0.01 \text{ kg food}} = \textbf{86,986 kg fish}$$

9. An additional 45.6% of TP will be removed by the microscreen so 1200 kg P in the discharge is equivalent to 6963 kg/year (19.1 kg/day) produced by the fish and the average allowable biomass is **190,759 kg fish.**

10. 500 L/min $= 8333$ cm^3/sec.   OFR $= 0.046$ cm/sec. $0.046 = 8333/A$, $A = 181,152$ cm$^2 = \textbf{18 m}^2$ **of clarifier surface required.**

Chapter 11

1. Assume 35% protein diet, $A = 19.6$ g TAN/kg food. Assume 3% BW/day feeding rate, 4000 kg fish = 80 kg food × 19.6 g TAN/kg food = 1.568 g TAN/day. Assume nitrification rate of 0.28 g/m² day,

$$\frac{1568 \text{ g TAN}}{\text{day}} \times \frac{\text{m}^2 \text{ day}}{0.28 \text{ g}} = 5600 \text{ m}^2 \text{ of biofilter surface required}$$

2. If disks are 2 m in diameter, each disk (2 sides) has 6.28 m² of surface area. 891 disks are required. If blocks are 2 m in diameter and 0.1 m thick, each one has 0.314 m³ of volume and 62.8 m² of surface area. **89 blocks are required.**

3. Fluidized bed filters contain media that averages 1313 m²/m³ and microbead filter media averages 2598 m²/m³. The required media volumes for fluidized bed and microbead filters are 4.27 m³ and 2.16 m³, respectively.

4. 100 mm Hg DO = 4.73 mg/L. 1.0 g TAN uses 4.34 g DO so 6805 g of DO are required per day.

$$\frac{6805 \text{ g}}{\text{day}} \times \frac{\text{day}}{1440 \text{ min}} \times \frac{1000 \text{ mg}}{\text{g}} \times \frac{\text{min}}{1000 \text{ L}} = 4.73 \text{ mg/L}$$

The biofilter will consume all the DO in the incoming water and, thus, will have to be aerated so that the biofilter effluent DO exceeds 2 mg/L.

5.

   a. 1.0 g TAN = 7.15 g alkalinity × 84/100 = 6.0 g $NaHCO_3$

   $$\frac{1568 \text{ g TAN}}{\text{day}} \times \frac{6 \text{ g NaHCO}_3}{\text{g TAN}} = \textbf{9408 g NaHCO}_3\textbf{/day}$$

   b. $\dfrac{0.11 \text{ kg NaHCO}_3}{\text{kg food}} \times \dfrac{80 \text{ kg food}}{\text{day}} = \textbf{8800 g NaHCO}_3\textbf{/day}$

   Use NO_2^- : Cl^- ratio of 1 : 3. 5.0 mg/L $NO_2^- = 15$ mg/L Cl^-

   $$\frac{15 \text{ mg Cl}^-}{\text{L}} \times \frac{111 \text{ CaCl}_2}{35.5 \text{ Cl}^-} = \textbf{47 mg/L CaCl}_2$$

6. Biofilter consumes 6.8 kg DO/day

$\log OC = 3.34 - 0.586 \ (\log 600), \ \log \ OC = 1.71, \ OC = 51.5 \ \text{mg DO/kg}$ fish hour.

$$\frac{51.5\,\text{mg}}{\text{kg hour}} \times 4000\,\text{kg} \times \frac{24\,\text{hour}}{\text{day}} \times \frac{\text{kg}}{10^6\,\text{mg}} = \frac{4.94\,\text{kg DO}}{\text{day}}$$

Total DO requirement is 6.81 + 4.94 = 11.75 kg DO/day. If absorption efficiency is 90%, the system will require **13 kg DO per day.**

7. $CC = \dfrac{(9.19 - 4.6)(0.0545)}{(3 \times 1.3 \times 0.529)/200} = \dfrac{24\,\text{lb}}{\text{gpm}} = \dfrac{2.89\,\text{kg}}{\text{L/min}} \times 10 = 28.9\,\text{kg/L/min}$

$$\frac{28.9\,\text{kg fish}}{\text{L/min}} \times \frac{0.01\,\text{kg food}}{\text{kg fish}} \times \frac{23.52\,\text{g TAN}}{\text{kg food}} \times \frac{1000\,\text{mg}}{\text{g}} \times \frac{\text{day}}{1440\,\text{min}}$$
$$= 4.72\,\text{mg TAN/L}$$

Allowable $NH_3 = 0.016$ mg/L, required f = 0.00339
$0.0039 = 1/(10^{9.564-pH} + 1)$ Solve for pH, **pH = 7.09**

8. A = 56(0.42) = 23.52 g TAN/kg food = 23,520 mg TAN/kg food
F = 0.00343, maximum allowable $NH_3 = 0.016$ mg/L, maximum allowable TAN = 4.67 mg/L

$$\frac{L}{4.67\,\text{mg}} \times \frac{23,520\,\text{mg}}{\text{kg food}} \times \frac{0.01\,\text{kg food}}{\text{kg fish}} \times 20,000\,\text{kg fish}$$
$$\times \frac{\text{day}}{1440\,\text{min}} = \textbf{700 L/min}$$

9. P = 0.42, A = 23.52 g TAN/kg food, f = 0.0034, $NH_3 = 0.80$ mg/kg food

$$20,000\,\text{kg fish} \times \frac{1.3\,\text{kg food}}{\text{kg fish}} \times \frac{0.80\,\text{mg NH}_3}{\text{kg food}} = \frac{20,800\,\text{mg NH}_3}{\text{day}}$$

Maximum $NH_3 = 0.016$ mg/L

$$\frac{20,800\,\text{mg NH}_3}{\text{day}} \times \frac{L}{0.016\,\text{mg}} = \frac{1.3 \times 10^6\,L}{\text{day}} = 903\ \text{L/min of new water required}$$

Chapter 15

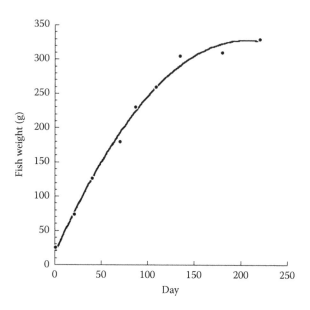

1. CSS when fish reach 235 g, 15,000 fish/3.5 Ha = **1007 kg/Ha**.
   CC when fish reach 337 g = **1444 kg/Ha**.
   With feeding, 500 g reached on day 163 = September 13.
   CC is still 1444 kg/Ha without feeding so 25,000 fish on November 1 would weigh 202 g each.
   10,000 fish on November 1 would weigh 505 g each.
   If 250 kg/Ha are removed on July 1, CC will be restored by November 1 and total production is **250 + 1440 = 1690 kg/Ha**.
2. 4000 fish/Ha × 0.350 kg/fish ÷ 0.5 kg = **2800 fish**.
3. Not fed: CSS at 163 g/fish = **651 kg/Ha**
   CC at 170 g/fish = **680 kg/Ha**
   Fed: CSS at 271 g/fish = **1083 kg/Ha**
   CC at 283 g/fish = **1327 kg/Ha**
4. $$1\,\text{Ha} \times \frac{10{,}000\,\text{m}^2}{\text{Ha}} \times 1\,\text{m} \times \frac{1000\,\text{L}}{\text{m}^2} \times \frac{5\,\text{mg}}{\text{L day}} \times 180\,\text{day} \times \frac{12\,\text{mg C}}{44\,\text{mg CO}_2}$$

   $$= 2.45 \times 10^9 \text{ mg C fixed by photosynthesis}$$

   200 kg fish × 25% dry matter × 50% C = 2.5×10^7 mg C in fish.
   1% of the C fixed in photosynthesis ended up in fish flesh.
5. 2000 kg fish × 25% dry weight × 75% protein × 16% N in protein = 60 kg N/Ha/year.

Chapter 16

1. DO : Density $= \dfrac{32,000\,mg}{mole} \times \dfrac{mole}{24.8\,L} = 1291.5\,mg/L$

$Ce = 1291.5 \times 0.0267 \times 0.20946 \times \dfrac{1}{1+(200/7600)} = \textbf{7.04 mg/L}$

$CO_2: Density = \dfrac{44,000\,mg}{mole} \times \dfrac{mole}{24.8\,L} = 1774.2\,mg/L$

$Ce = 1774.2 \times 0.6810 \times 0.0004 \times \dfrac{1}{1+(200/7600)} = \textbf{0.47 mg/L}$

2. DO: $[1295.5 \times 0.02711 \times 0.20946] - [17\ (0.0841 - 0.00256(28) + 0.0000374(28^2)] = \textbf{6.65 mg/L}$
   CO_2: $[1781.3 \times 0.6986 \times 0.0004] - [(0.062-0.004\ (28) + 0.00004(28^2) + 0.002\ (17)] = \textbf{0.48 mg/L}$
3. There is still a trace of CO_2 at pH values greater than 8.3, but photosynthesis is mostly fueled by the carbonic anhydrase conversion of HCO_3^- to CO_2.
4. f at pH 6 = 0.0007, NH_3 = **0.0007 mg/L**
   f at pH 9 = 0.412, NH_3 = **0.412 mg/L**
5. f at pH 10 = 0.798 × 0.3 = **0.239 mg/L NH_3**

Chapter 17

1. $\dfrac{[H^+]\left[HCO_3^-\right]}{[CO_2]} = 4.47 \times 10^{-7}$

$\dfrac{100\,mg\ HCO_3^-}{L} \times \dfrac{mole}{61,000\,mg} = 1.64 \times 10^{-3}\,M$

$\dfrac{0.5\,mg\ CO_2}{L} \times \dfrac{mole}{44,000\,mg} = 1.14 \times 10^{-5}\,M$

$\dfrac{[H^+][1.64 \times 10^{-3}]}{[1.14 \times 10^{-5}]} = 4.47 \times 10^{-7}$ Solve for $[H^+]$

$[H^+] = 3 \times 10^{-9}$, **pH = 8.51**

2. $100 \, \text{mg HCO}_3^-/\text{L} \times \dfrac{100 \, \text{formula weight of CaCO}_3}{61 \, \text{formula weight of HCO}_3^-} = \mathbf{164 \, mg/L \ CaCO_3}$

3. $\dfrac{[10^{-8.2}]\left[\text{HCO}_3^-\right]}{[1.14 \times 10^{-4}]} = 4.47 \times 10^{-7}$

$\text{HCO}_3^- = 0.0081 \, \text{M} = \mathbf{493 \, mg/L} \times \dfrac{100 \, \text{CaCO}_3}{61 \, \text{HCO}_3^-} \ \textbf{Alkalinity} = \mathbf{808 \, mg/L \ CaCO_3}$

4. $2116 \div 179\% = \mathbf{1182 \ kg/Ha}$ for CaO. For dolomite, $2116 \div 109\% = \mathbf{1941 \ kg/Ha}$

5. $\dfrac{5.0 \, \text{mg}}{\text{L}} \times \dfrac{\text{kg}}{10^6 \, \text{mg}} \times \dfrac{1000 \, \text{L}}{\text{m}^3} \times 10{,}000 \, \text{m}^3 = 50 \, \text{kg in water}$

   1950 kg in mud

6. $\dfrac{\left[\text{CO}_3^{2-}\right][\text{H}^+]}{\left[\text{HCO}_3^-\right]} = 4.68 \times 10^{-11}$

   $\dfrac{\left[\text{CO}_3^{2-}\right][10^{-6}]}{[1.64 \times 10^{-3}]} = 4.68 \times 10^{-11}, \left[\text{CO}_3^{2-}\right]$

   $= 7.7 \times 10^{-8} \, \text{M}, \ \text{CO}_3^{2-} = \mathbf{4.6 \times 10^{-3} \, mg/L}$

7. $\dfrac{\left[\text{CO}_3^{2-}\right][10^{-10}]}{[1.64 \times 10^{-3}]} = 4.68 \times 10^{-11}, \left[\text{CO}_3^{2-}\right] = 7.7 \times 10^{-4} \, \text{M}, \ \text{CO}_3^{2-} = \mathbf{46 \, mg/L}$

8. $100 \, \text{mg/L CaCO}_3 = 0.00164 \, \text{M HCO}_3^-$
   $10 \, \text{mg/L CO}_2 = 0.00023 \, \text{M CO}_2$
   pH = **7.2**
   If $\text{CO}_2 = 5 \, \text{mg/L}$ (0.000114 M), pH = **7.5**
   If $\text{CO}_2 = 2.5 \, \text{mg/L}$ (5.68×10^{-5} M), pH = **7.8**

9. $15 \, \text{mg CaSO}_4/\text{L} \times 100 \ (\text{CaCO}_3)/136 \ (\text{CaSO}_4) = 11 \, \text{mg/L} + 50 \, \text{mg/L} = \mathbf{61 \ mg/L}$. The alkalinity will not change.

10. $0.8 \times 5600 = \mathbf{4480 \ kg/Ha}$

Chapter 18

1. 100 kg straw × 40% C × 7% = 2.8 kg bacterial C = 5.6 kg bacterial biomass × 10% = 0.56 kg N needed. Straw contains 0.1 kg N so **0.46 kg N** will be immobilized from the soil.

2. 0.56 kg N needed, 0.1 kg present, 1.0 kg added, **0.54 kg** will be mineralized to the soil.

3. 1000 kg × 45% × 8% = 36 kg bacterial C = 72 kg bacterial biomass = 7.2 kg N needed. Manure contains 22 kg N so **14.8 kg N** will be mineralized.
4. Phosphorus is usually the limiting factor in aquatic photosynthesis and, thus, promotes the production of *Cyanobacteria* which fix atmospheric N.
5. Superphosphate contains 20% P_2O_5 (0-20-0). 10.5 Ha × 50 kg/Ha ÷ 20% = **2625 kg**.
6. Use ammonium polyphosphate (10-34-0) to satisfy the P requirement. The density is 1.2 kg/L. 45 kg/0.34 ÷ 1.2 kg/L = 110 L ammonium polyphosphate. 110 L ammonium polyphosphate contains 13.2 kg N and 25 kg (11.8 more) are required. Make up the difference with Nitan-Plus (29-0-0). 40.7 kg or **34 L of Nitan Plus is required**.
7. Fish production responds to fertilizer input following the Law of Diminishing Returns. Each fertilizer input (22.4 kg P) is worth $46.67. The value of the first input is $317 − $97 = $220.

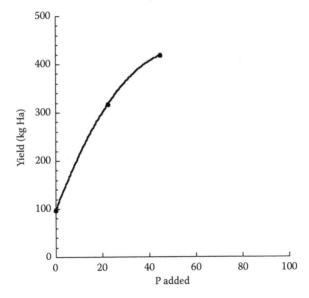

The value of the second input is $418 − $317 = $101. The value of a projected third input is about $437 − $418 = $19. Thus, slightly more than two units of input provide the best cost benefit.
8. Mix triple superphosphate (0-46-0) and urea (45-0-0). For 100 kg fertilizer, 20 kg P_2O_5 or 43.5 kg triple superphosphate is required. 20 kg N = 44.4 kg urea. Add 12.1 kg $CaCO_3$ to bring total weight to 100 kg.
9. Add urea. 200 kg fertilizer will contain 40 kg N. 100 kg of monoammonium phosphate contains 11 kg N so 29 kg more is needed.

29/.45 = 0.64 kg urea. Recipe for 200 kg of mixed fertilizer is **100 kg monoammonium phosphate + 64 kg urea + 36 kg CaCO$_3$.**

10. $\dfrac{30\,\mu g}{L} \times 5 \times 10^6\, L \times \dfrac{kg}{10^9\,\mu g} = 0.15\,kg\ P = 0.64\,kg$ Phosphoric acid $\times \dfrac{L}{1.2\,kg}$

$$= 0.53\,L\ \textbf{Phosphoric acid}$$

3 kg N, 10.3 kg Nitan Plus, **8.6 L Nitan Plus required.**

Chapter 19

1. DF = −0.024 DO + 2.677
   12 mg/L at 30°C = 159% saturation, DF = **−1.14 mg/L**
   14 mg/L at 20°C = 158% saturation, DF = −1.12 mg/L

2. log O$_2$ = −0.999−0.000957 (300) + 0.0000006 (300^2) + 0.0327 (29) − 0.0000087 (29^2) + 0.0000003 (300 × 29)
   log O$_2$ = −0.289, O$_2$ = 0.515 mg/g hr

$$\dfrac{0.515\,mg}{g\ hour} \times 12\ hour \times \dfrac{1000\,g}{kg} \times 18{,}000\,kg$$

$$= \dfrac{341{,}280{,}000\,mg}{120{,}000\,m^3} \times \dfrac{m^3}{1000\,L} = 0.93\,mg/L$$

Remove 9000 kg (30,000 fish), 30,000 fish remain and grow to 600 g each,
log O$_2$ = −0.411, O$_2$ = 0.388 mg/g hour = 0.70 mg/L

3. BOD = −1.133 + 0.00381 S + 0.0000145 S^2 + 0.0812 T − 0.000749 T^2 − 0.000349 ST
   April: BOD = 0.069 mg/L hour × 12 hour = **0.83 mg/L**
   July: BOD = 0.46 mg/L hour × 12 hour = **5.52 mg/L**

4. DF = −0.024 (180.6) + 2.677 = −1.66 mg/L

$$DO_{fish} = \dfrac{0.448\,mg}{g\ hour} \times 12\ hour \times \dfrac{1000\,g}{kg} \times \dfrac{2250\,kg}{Ha}$$

$$\times \dfrac{Ha}{10{,}000\,m^3} \times \dfrac{m^3}{1000\,L} = 1.21\,mg/L$$

$$DO_{mud} = \dfrac{61\,mg}{m^2\ hour} \times 12\ hour \times \dfrac{10{,}000\,m^2}{Ha} \times \dfrac{Ha}{10^{7}L} \times = 0.73\,mg/L$$

BOD = 0.44 mg/L hour × 12 hour = 5.29 mg/L
DO$_{dawn}$ = 14 − 1.66 − 1.21 − 0.73 − 5.29 = **5.11 mg/L**

5. The DO is predicted to be 2.26 mg/L at sunrise so emergency aeration will probably not be required.

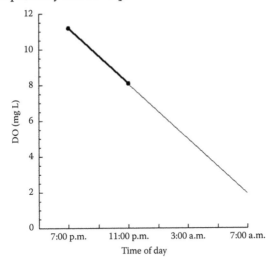

Index

Milton Keynes UK
Ingram Content Group UK Ltd.
UKHW040444071024
449327UK00020B/974